Environment and Society

At the start of the twenty-first century, it can be argued that human societies have a greater impact on the environment than ever before. We have always been dependent upon, and interacted with, the 'natural' environment. However, the dramatic social changes of the past three centuries have altered the form of our relationship with non-human nature to the extent that some would see people–planet relations as in a situation of crisis.

Environment and Society is an introduction to the sociological study of the environment. It emphasizes the ways in which our conceptualization of the relationships between environments and human societies differ historically and cross-culturally. Beginning with an overview of the different ways contemporary sociologists have theorized the relationship between societies and their environments, the book then considers how the main strands of 'green' social and political theory have understood environment–society relations and why they see the current forms of these relationships as problematic. Particular social processes and practices such as industrialization/post-industrialization, urbanization, 'development' and globalization, are then examined in terms of their impact on the environment. Contesting perspectives on the relationship between the 'social' and the 'natural' draw the book to a close, with a consideration of human interactions with animals.

Environment and Society provides a comprehensive and critical account of the ways in which we can think about the relationship between human societies and the environments with which they interact. It argues that human societies are ecologically embedded, and that environments are also often socially embedded and constituted. It makes the different theoretical positions and empirical studies accessible to students, and includes chapter outlines and summaries, annotated further reading, boxed case studies and discussion points.

Erika Cudworth is a Senior Lecturer in Sociology and Politics at the University of East London.

Routledge Introductions to Environment Series
Published and Forthcoming Titles

Titles under Series Editors:
Rita Gardner and A.M. Mannion

Environmental Science texts

Atmospheric Processes and Systems
Natural Environmental Change
Biodiversity and Conservation
Ecosystems
Environmental Biology
Using Statistics to Understand the
   Environment
Coastal Systems
Environmental Physics

Forthcoming:
Environmental Chemistry
   (December 2002)

Titles under Series Editor:
David Pepper

Environment and Society texts

Environment and Philosophy
Environment and Social Theory
Energy, Society and Environment
Environment and Tourism
Gender and Environment
Environment and Business
Environment and Politics (2nd edition)
Environment and Law
Environment and Society

Environmental Policy (December 2002)
Environmental Movements
   (March 2003)
Environmental Values (April 2003)
Representing the Environment
   (June 2003)
Environment and the City
   (January 2004)
Environment and Sustainable
   Development (December 2004)

Routledge Introductions to Environment Series

Environment and Society

Erika Cudworth

Routledge
Taylor & Francis Group

LONDON AND NEW YORK

First published 2003 by Routledge
11 New Fetter Lane, London EC4P 4EE

Simultaneously published in the USA and Canada
by Routledge
29 West 35th Street, New York, NY 10001

Routledge is an imprint of the Taylor & Francis Group

© 2003 Erika Cudworth

Typeset in Times by Keystroke, Jacaranda Lodge, Wolverhampton
Printed and bound in Great Britain by TJ International, Padstow, Cornwall

British Library Cataloguing in Publication Data
A catalogue record for this book is available from the British Library

Library of Congress Cataloging in Publication Data
Cudworth, Erika, 1966–
    Environment and society / Erika Cudworth.
      p. cm. – (Routledge introductions to environment)
    Includes bibliographical references and index.
    1. Human ecology. 2. Nature–Effect of human beings on.
    3. Environmental protection. I. Title. II. Routledge introductions to
    environment series

    GF41 C83 2002
    304.2–dc21   2002068236

ISBN 0–415–21617–6 (hbk)
ISBN 0–415–21618–4 (pbk)

Contents

Series editor's preface
Environment and Society titles

The modern environmentalist movement grew hugely in the last third of the twentieth century. It reflected popular and academic concerns about the local and global degradation of the physical environment which was increasingly being documented by scientists (and which is the subject of the companion series to this, *Environmental Science*). However it soon became clear that reversing such degradation was not merely a technical and managerial matter: merely knowing about environmental problems did not of itself guarantee that governments, businesses or individuals would do anything about them. It is now acknowledged that a critical understanding of socio-economic, political and cultural processes and structures is central in understanding environmental problems and establishing environmentally sustainable development. Hence the maturing of environmentalism has been marked by prolific scholarship in the social sciences and humanities, exploring the complexity of society–environment relationships.

Such scholarship has been reflected in a proliferation of associated courses at undergraduate level. Many are taught within the 'modular' or equivalent organizational frameworks which have been widely adopted in higher education. These frameworks offer the advantages of flexible undergraduate programmes, but they also mean that knowledge may become segmented, and student learning pathways may arrange knowledge segments in a variety of sequences – often reflecting the individual requirements and backgrounds of each student rather than more traditional discipline-bound ways of arranging learning.

The volumes in this *Environment and Society* series of textbooks mirror this higher educational context, increasingly encountered in the early twenty-first century. They provide short, topic-centred texts on social science and humanities subjects relevant to contemporary society–environment relations. Their content and approach reflect the fact that each will be read by students from various disciplinary backgrounds, taking in not only social sciences and humanities but others such as physical and natural sciences. Such a readership is not always familiar with the disciplinary background to a topic, neither are readers necessarily going on to further develop their interest in the topic. Additionally, they cannot all automatically be thought of as having reached a similar stage in their studies – they may be first-, second- or third-year students.

The authors and editors of this series are mainly established teachers in higher education. Finding that more traditional integrated environmental studies and specialized texts do not always meet their own students' requirements, they have often had to write course materials more appropriate to the needs of the flexible undergraduate programme. Many of the volumes in this series represent in modified form the fruits of such labours, which all students can now share.

Much of the integrity and distinctiveness of the *Environment and Society* titles derives from their characteristic approach. To achieve the right mix of flexibility, breadth and depth, each volume is designed to create maximum accessibility to readers from a variety of backgrounds and attainment. Each leads into its topic by giving some necessary basic grounding, and leaves it usually by pointing towards areas for further potential development and study. There is introduction to the real-world context of the text's main topic, and to the basic concepts and questions in social sciences/humanities which are most relevant. At the core of the text is some exploration of the main issues. Although limitations are imposed here by the need to retain a book length and format affordable to students, some care is taken to indicate how the themes and issues presented may become more complicated, and to refer to the cognate issues and concepts that would need to be explored to gain deeper understanding. Annotated reading lists, case studies, overview diagrams, summary charts and self-check questions and exercises are among the pedagogic devices which we try to encourage our authors to use, to maximize the 'student friendliness' of these books.

Hence we hope that these concise volumes provide sufficient depth to maintain the interest of students with relevant backgrounds. At the same time, we try to ensure that they sketch out basic concepts and map their territory in a stimulating and approachable way for students to whom the whole area is new. Hopefully, the list of *Environment and Society* titles will provide modular and other students with an unparalleled range of perspectives on society–environment problems: one which should also be useful to students at both postgraduate and pre-higher education levels.

David Pepper

May 2000

Series International Advisory Board

Australasia: Dr P. Curson and Dr P. Mitchell, Macquarie University

North America: Professor L. Lewis, Clark University; Professor L. Rubinoff, Trent University

Europe: Professor P. Glasbergen, University of Utrecht; Professor van Dam-Mieras, Open University, The Netherlands

Acknowledgements

I dedicate this book to my parents, Jill and Tony, for all their love and support, and to Dave, for the same, and for culinary delights and exemplary single parenting when the manuscript was due. Apologies to baby Jake for being an often rather distracted mother during his first year, and (oh so) very many thanks for sleeping through the night from seven weeks.

I am also grateful to the non-human members of our household, without whom the business of writing would be far more isolating. Thanks to Mooki, Phoebe and Thumbelina the felicitous felines for their quiet affection, and to Kevin (?) the Jack Russell, for keeping me company from under the desk and not farting too much.

My thanks to the third-year politics undergraduates at the University of East London, and to my friend and colleague John McGovern, for reading and commenting on Chapter 5. The contents of Chapters 1 and 2 have been read in earlier incarnations by Michael Rustin at the University of East London, and Sylvia Walby at the University of Leeds, and the material is much improved as a result of their suggestions. I am particularly grateful to Judith Burnett both for her thoughtful and stimulating comments on Chapter 6, and for helping me hang on to my sense of humour throughout our unending encounters with 'quality assurance' in higher education.

Various friends and relations helped in providing some of the illustrative material – Tony Cudworth (Figures 4.4 and 4.5), John Monger (Figure 6.5) and thanks in particular to Paul Goodey (Figures 4.2 and 4.3). The League Against Cruel Sports (Figure 4.1) and Compassion in World Farming (Figures 6.1, 6.2, 6.3 and 6.4) were kind enough to allow me to use the photographs in figures in Chapters 4 and 6 without charge, my thanks to Claire Thomas and Kathy Dilks respectively. The World Wide

Fund for Nature allowed me use the photograph in Figure 5.1 – thanks to Patricia Patton for advice, and for searching out some images for me. Robin Webb from Vegetarian Shoes of Brighton generously 'donated' the marvellous cow in recycled rubber boots in Figure 3.1.

I would also like to thank Ann Michael at Routledge, for her guidance on preparing the manuscript and the three anonymous reviewers for their helpful comments on the first draft.

 **Introduction**

What is the relationship between 'nature' and society? What kinds of impact do human groups make upon the planet? How have environmental (or ecological) limits shaped human behaviour, cultural practices and social institutions? What do developments in science and technology, economic practice and government policy tell us about the changing forms of nature–society relations? How are these changes represented through the media and popular culture? These are all key questions for environmental sociology, and for this book. These are also questions that are unlikely to become less significant in the near future.

This paragraph is one of the last to be written in the history of this book, and as I write it, I am reminded of the pertinence of the questions above. The BBC Breakfast News announced this morning, 15 February 2002, that 'scientists in Texas' have successfully cloned a domestic cat. Scientists parade a 'cute' kitten for the media, and repeatedly attempt to get this techno-beast to sit in a glass laboratory jar for the photographs. 'Many American pet owners' are apparently keen to have their favourite fluffy friends exactly reproduced. Yesterday, we were informed that the regional Scottish Assembly had banned fox hunting. These developments raise questions about relations between humans and other species, the appropriate use of genetic technology, the relationship between 'science' and society, the use of the 'countryside', the effect of lobbying and direct political action.

This book is an introduction to the sociological study of the interrelationships between environments and human societies. It emphasizes the ways in which our conceptualization of the relations between society and the environment differs historically and cross-culturally, and examines the ways sociologists have attempted to theorize and to empirically investigate environment–society relations. However, before we consider what sociology might have to offer to the study of the environment, and how it might help us understand some of the issues mentioned above, we need to think a little about the 'environment' and 'society'.

The term 'environment' can mean almost anything: a street, an aeroplane, a lecture hall or a forest. Being concerned with the equally general 'study of society', sociologists have tended to define the environment as a human manufacture, for example, they have been concerned with the 'built environment' of the city, its social characteristics and impact on cultural norms and values. The development of environmental science, particularly of ecology, has led to the more general understanding of the environment that is now part of everyday knowledge. We can define the environment as the context which provides the conditions for the existence of the human species, and for the multifarious other living creatures and vegetation. For the ecological scientist, there are webs of interconnecting relations of dependency and reciprocity between humans, other animal life, plants, water and earth.

A key spur to academic interest in the environment in the past thirty years, has been scientific research revealing sometimes alarming patterns of global environmental change. Such findings have been well publicized by the media, at least in Western states. Such environmental hazards include global warming, deforestation, resource depletion, pollution, damage to the ozone layer, the erosion and desertification of land. The environment, until recently, has often been seen as an area of inquiry requiring expertise in the natural sciences. The aforementioned 'environmental problems' involve trees, the atmosphere, marine life, etc. – all things which social scientists have often assumed scientists know about, and social scientists cannot. Scientists have attributed the scale of environmental hazard largely to human activity. The scientific and technological advances of modern and modernizing societies, with their varied attempts to harness natural powers, have not been without deleterious environmental effects. These include the global warming resultant from pollution by-products of certain methods of electricity production, or the creation of toxic wastes by the generation of nuclear power. In addition, the spread of industrial production and urban living impose burdens on natural resources. Alongside media attention to such issues from the 1970s onwards, there was the development of environmental protest movements which sociologists of social movements were compelled to take note of.

While academic study of the environment within the social sciences has increased dramatically in recent decades, concern with environment–

society relations is not necessarily a new phenomenon. This book argues that environment–society relations are not a set of 'new' questions for sociologists, as an examination of some of the key issues concerning sociologists throughout the nineteenth and twentieth centuries (such as urbanization, industrialization, development and modernization) will show. However, until twenty years ago, environmental issues were implicit in sociological studies rather than explicit. In the 1990s, environmental sociology became a small but recognized sub-field of the discipline, reflecting, in no small part, the increased media coverage of environmental issues and expansion in political activism by environmental social movements. A substantial part of the content of the book is of relatively recent origin, and this reflects the intensification of sociologists' interest in the environment.

The dramatic increase in social science material on the environment since the early 1990s means that there are inevitable overlaps of themes, theories and issues across texts. Given the relative historical absence of material covering this area however, and the potential scope of coverage, each of the recent texts will have its own preoccupations and specific focus. There are other books in this series to which readers can usefully refer, to further their reading and deepen their understanding of some of the issues covered here. For example, material on environmental social movements is covered by Chapter 3 of this book and is also a feature of *Environment and Politics* (Doyle and McEachern 1998). The material covered in the 'politics' text covers the organization and structure of environmental movements in some detail, whereas this chapter focuses on the social conditions contributing to the development of environmental movements, and the social bases of their support. *Environment and Politics* also has a chapter on global dimensions of environmental politics, which focuses on policy-making and supra-governmental conferences and initiatives. The chapter on globalization in this text examines the cultural and economic aspects of globalization. It is more concerned with theoretical debates and the usefulness of globalization theory for making sense of the very different environmental problems in wealthy and poorer societies of the globe.

Another text in this series *Environment and Social Theory* (Barry 1999) is, as its title suggests, a theoretical text whereas *Environment and Society* is also concerned with the usefulness of empirical sociological studies of environment–society relations. *Environment and Social Theory* covers the social sciences more broadly, including philosophy, psycho-social theory, economic and political theory. The theoretical focus in this book is on contemporary debates in sociology. In *Environment and Social Theory*

(1999), John Barry covers, in far greater detail, the theoretical debates on scientific knowledge and the concept of 'risk'. Those readers looking for detailed coverage of important contemporary contributions, such as those of Ulrich Beck and Jürgen Habermas, for example, should refer to Barry's contribution to this series. In his recent *Sociology and the Environment*, Alan Irwin (2001) also concentrates on science and environmental risk as a framework for discussing contemporary theoretical debates and policy orientations. *Environment and Society* focuses on the historical development of environmental concern within sociology, and indicates what 'classical sociology' may have to contribute. While questions of the relationship between the social and natural sciences, and of the status of scientific knowledge are examined by this book, they are not its main focus. In many ways, this text is more of an overview of the disciplinary contribution of sociology than that of Irwin, or Michael Bell's (1998) more introductory *An Invitation to Environmental Sociology*. One of the intentions of this book is to examine some areas of sociological interest that have not been covered by other texts. As such, it will consider relations between animals and societies, the impact of social processes such as industrialization/de-industrialization, urbanization and the particularly contested fields of 'development' and globalization, on environments.

This book has a number of overall aims which are:

- to outline the historical contribution which sociology as a discipline has made to our understanding of the environment;
- to critically discuss the theoretical debates in contemporary sociological theory and empirical research;
- to explore sociological understandings of environment–society relations in the context of particular environmental issues and problems;
- to examine the ways different schools of 'environmentalist' thought conceptualize the relationship between human societies and 'nature';
- to consider the ways in which social movements have challenged contemporary social practices and institutional arrangements which they see as damaging to the environment.

Outline of the book

The book has six chapters. The first two examine different ways of theorizing relations between 'society' and the 'environment', and the last four examine particular issues in environmental sociology and combine some theory with a discussion of empirical research on various topics.

Chapter 1 begins with an examination of how the discipline of sociology developed on the basis of drawing a firm distinction between the 'social' world and the 'natural' world. This meant that 'sociology' came to be defined as the study of 'society', and this society was conceptualized as the relationships between human beings. The chapter considers how this might help to explain the reticence of most sociologists (at least until the late 1980s and early 1990s) to study the environment. The chapter presents an outline of the different ways in which contemporary sociologists have attempted to theorize relations between human society and the environment, and suggests how future environmental sociologies might develop. It will argue that studying the environment presents a challenge for the discipline and suggest that, while some environmental sociologies will continue operate within the established analytical frameworks of the discipline, others are likely to try and reformulate them.

In sociology, there have been a number of long-standing and highly contentious disputes surrounding the theorization of relations between the environment and society. These can be described, albeit rather crudely, as realism, constructionism and co-constructionism. These positions tap into long-standing theoretical questions in the discipline more widely. One such question concerns the relationship between social structures (social institutions and practices which affect our behaviour) and agency (our ability to make choices and direct the course of our lives, or more generally, the power we have to change things). For example, some sociologists see both the human and the natural world as consisting of a series of structures (processes and relationships). These structures shape how human beings behave, and how we treat our environment. Others are keener to emphasize how different groups of humans ('agents') see the environment and act on it in different ways. Second, there is disagreement in social theory between 'realists' and 'interpretivists' (or 'social constructionists'). 'Realists' argue there is a real environment 'out there', which exists independently of how we humans think about it. Social constructionists, on the other hand, tend to argue that 'the environment' or 'nature' means different things to different groups of people depending on the particular norms and values of a society, that is, the environment is what we humans think it is. Co-constructionists argue that we construct 'reality' and it in turn constructs us. These positions are simply put here, and there is a tendency for adversaries to caricature opponents' arguments in order to discredit them. My own view is that whatever the 'position' articulated, constructionists and realists mostly accept both that a material environment 'exists', and also that it is understood and represented in a variety of ways. They dispute the relative significance of material (physical, institutional, economic) compared with the ideological (cultural, symbolic, representational) factors in the constitution of social life however. These issues deserve some discussion so early in this book because they will reappear in some form in each of the chapters that follow.

Chapter 2 shifts focus from looking at how sociologists have conceptualized the relationship between environment and society, to looking at how theorists within the 'green' or environmental movement have understood such a relationship. The chapter outlines, compares and evaluates four strands of 'green' social and political theory: social ecology, eco-socialism, deep ecology and eco-feminism. The chapter concentrates on the ways in which these different kinds of green theory have analysed issues that are of importance to sociologists, such as social inclusions and exclusions based around class, 'race' and gender.

The third chapter considers the ways environmental social movements have challenged environmental problems and issues. The chapter begins by looking at how sociologists have attempted to explain the development of social movements in general and the environmental movement in particular. It looks at a range of strategies and tactics employed by differing organizations within the environmental movement, and considers their effectiveness in influencing public opinion, political parties, government policy and the media representation of environmental issues and of environmental activists.

The next three chapters look in detail at particular issues that have been identified by environmental social movements and 'green' theorists. These issues are: Western industrialism and urbanization, the global impact of such 'development' in the poorer societies of the world, and our relationship to other (i.e. non-human) animals. The chapters will show how concepts, theories and empirical studies from sociology may help our understanding of the causes and consequences of environmental issues.

Chapter 4 examines how sociologists have understood the processes of urbanization and industrialism in the development of modern Europe and North America, and links such understandings to contemporary environmental problems. Some of the ideas of 'classical' and contemporary sociologists on economic development, work and industrial production are examined, and considered in relation to problems such as pollution and global warming. The chapter looks at urban/rural divisions and the different 'ways of life' which sociologists have claimed characterize rural and urban societies. The implications of such studies are relevant to environmental issues associated with urban living such as population over-crowding and traffic congestion. The chapter also looks at recent sociological work on the rural environment, and considers concepts of the 'countryside' and 'wilderness' alongside those of the town and city, and the issues of intensive chemical and biotechnical agricultural production.

The fifth chapter focuses on the contribution of sociology to our understanding of 'global' environmental problems. It discusses the concept of a 'global environmental system', concentrating in particular on the processes of development in relation to globalization. The chapter examines contemporary theories and perspectives on the globalization process, in sociology generally, and with reference to global environmental change specifically. It considers the role of transnational corporations, supra-national political institutions, and the spread of 'Western' consumerism in changing social and environmental features of the poorer regions of the globe. It examines issues such as agricultural production, corporate bio-technology, international trade and debt, pollution and resource depletion, and food shortages, and gives an overview of how these problems may be understood using concepts from the sociology of globalization and development.

The final chapter returns to one of the opening themes of the book, the relationship between the 'natural' and the 'cultural'. It looks at the different approaches sociologists have adopted in attempting to understand the relationship between humans and non-human animals. The chapter begins by contrasting sociological perspectives on the changing relations between humans and non-human animals, both historical and contemporary. It proceeds to examine and critique the ethical debate surrounding 'rights' for non-human animals, and looks at the politics of animal welfare using Britain as an illustrative example. Finally, it considers a number of issues in contemporary Western societies, which are key arenas of human–animal relations: farming, pet-keeping and the use of animals for human entertainment (through zoos or hunting for example).

The book ends with a brief conclusion that sums up the key arguments of the six chapters. It re-emphasizes the usefulness of certain sociological concepts and theories in enhancing our understanding of the relationship between the environment and the societies which inhabit and interrelate with it in focusing on the issue of genetic modification of animals, plants and possibly of humans.

Further reading

A readable introduction to environmental problems and issues from a sociological perspective is Steven Yearley's *The Green Case: A Sociology of Environmental Issues, Arguments and Politics* (London: Routledge, 1992).

More recent, is Alan Irwin's *Sociology and the Environment* (Cambridge: Polity, 2001), which focuses on particular sites of environment–society relations, such as the sociology of scientific knowledge, the sociology of risk, policy-making and new technologies.

Very accessible is Michael M. Bell's *An Invitation to Environmental Sociology* (Thousand Oaks/London: Pine Forge/Sage, 1998). This is more introductory than the Irwin text and, given its activist perspective, is a lively read.

1 Sociology and the environment

This chapter will:

- introduce different sociological perspectives
- explore the ways in which sociology is important to our understanding of the environment
- identify and critically discuss different approaches in environmental sociology, including social constructionism, critical realism and actor-network theory
- consider the possible future(s) of environmental sociology

Introduction

This chapter examines what sociology may be able to contribute to the study of the environment. Sociology is an incredibly broad discipline within the social sciences, and can be defined at the most general level as the study of human social life. It looks at our behaviour as social beings, our ideas and 'values' and how these differ across historical time and cultural space. It examines the social networks and institutions which shape our lives, and which we change by our actions, such as the household, the education system and the many differing kinds of workplace. It considers how we interact with each other and helps us to unpick and question what might be seen as everyday phenomena: eating a meal, having sex, watching television, worrying about the time. Sociology has also been concerned with differences and divisions within societies such as class, caste, ethnicity, 'race', gender, age and sexuality. In 'being about' society however, sociology has not seen itself as concerned with 'nature' and for most of its history, the environment has not been seen as 'social' and not considered within the scope of the discipline.

The situation is slowly changing, and the sociology of the environment is currently a small, but rapidly developing area. This chapter begins by asking why sociologists have been reticent about studying the environment. It looks at the historical legacy of the discipline which drew a firm boundary between what was 'natural' and assumed to be the subject matter of the 'natural sciences', and what was cultural or social.

The chapter proceeds to look at some examples of the contemporary theoretical perspectives that have emerged in sociology as a means of explaining relations between the 'environment' and 'society'. **Social constructionism**, **critical realism** and **co-constructionism** are outlined in some detail in this chapter and will be referred to throughout the book. The heated exchanges between adherents of differing sociological approaches have been called a 'sterile debate' (Wynne 1996: 44), or a 'rather dull debate' (Macnaghten and Urry 1998: 2). Phil Macnaghten and John Urry rather grandly claim to propose a way of thinking about society–'nature' relations which 'transcend(s)' such divisions. However, what they do is to caricature these positions and then outline their own, which is firmly on the social constructionist side of this divide. There does seem to be a tendency for each 'side' to caricature its opponents as a means of making its criticisms self-evident. Kate Burningham and Geoff Cooper (1999) suggest the differences between each 'camp' are not as marked as the protagonists often assume, but nevertheless, their key point is 'to emphasize the appeal and utility of social constructionism' (1999: 297). Peter Dickens (2001) has advocated the use of critical realism in understanding relations between the social and the 'natural', and argued that this can provide the basis for the development of a 'single science'. These schools of thought are vital to contemporary analyses of the environment and society and are more often lively than dull, for, as Dickens suggests, 'in so far as academia is capable of having a stand-up row, it is over this issue that sociology has become most heated' (1996: 72). The chapter will close by considering what these approaches contribute to our understanding of environment–society relations, what issues sociologists have taken an interest in so far, and what they might look at in the near future. Before embarking on any of this however, it would be helpful to have a quick look at some of the long-standing controversies in sociology to which environmental sociologists refer.

Sociological perspectives

One key controversy is the relationship between 'society' and the individual. Some sociologists have seen the individual human being or collectivity of human beings as key to the constitution of social life, arguing that social institutions and practices are the result of the actions of individuals. These approaches stress the significance of human action or '**agency**' in shaping social life. Others however, stress the significance of social '**structures**', that is, of social institutions such as the state, the household or organized religion, and the social processes that make up these institutions (e.g. getting married, attending a religious service), as most significant in constituting 'society'.

There are other important sociological disputes between 'macro' and 'micro' approaches, and between those emphasizing the experience of the 'individual' and those who stress 'the social'. Macro approaches concentrate on larger-scale features of society such as institutions, organizations and culture, whereas micro approaches look at the more personal aspects of social life such as day-to-day personal encounters, emotional life and personal experiences. Derek Layder (1994: 2–3) calls the divisions between agency/structure, individual/social and macro/micro the 'three key dualisms' in sociology. He argues that many sociological approaches do not sit fast on one side of a divide, but take on board aspects of, for example, microsociology within a generally macro approach. The difference lies in the emphasis or degree of importance a sociologist gives different levels of analysis or different aspects of social life.

Related to the three dualisms described above are the **ontological** and **epistemological** differences between those who subscribe to **postmodernism** as a series of approaches in sociology, and those who retain some form of what we can call 'modernist' analysis. The term 'postmodernism' itself can mean a variety of things, from an architectural style to a method of literary criticism. In terms of sociology, it has meant both a form of theorizing and a conception of society itself as characterized by fragmentation and uncertainty, and has become increasingly influential since the 1980s. '**Modernism**' can be used to describe the social science theories and perspectives that developed from the eighteenth century. These attempted to explain, in non-religious rationalist terms, the dramatic processes of change that have come to be known as the transitions to modernity. Beginning in Europe from the fifteenth century, these include: **the 'Enlightenment'** and the development of rationalist and scientific thought, the development of capitalism and the changes it produced in agricultural production, the development of industrialism and the process of urbanization.

Modernist theories sought causal explanations of social processes. Questions to which nineteenth-century sociologists sought answers included: why capitalism emerged as a form of social and economic organization, the role of religion in the process of modernization, why was there was mass migration of populations to towns and cities, what effect did urbanization have, on social norms, values and behaviour? The kinds of explanations of social life produced by sociologists of the nineteenth and much of the twentieth centuries are seen by theorists of postmodernity to be characterized by 'grand narratives', overarching ways of seeing and understanding the world (Lyotard 1984; Lash 1990). Such narratives include 'worldviews' about progress, science and rationality, and overarching theories which contest the institutions and practices of modernity, for example, Marxist theories about the operation of capitalism and feminist theories of male domination or 'patriarchy'. The era of postmodernism is seen as characterized by social diversity and the fragmentation and difference of people's experiences of social life. The postmodern world is an 'uncertain' one (Lash and Urry 1994: 257), and sociologists influenced by postmodern thinking have turned away from a concern with large-scale social processes to focus on more 'micro' sociological concerns such as people's subjective experiences.

Sociologists who adopt a postmodern approach tend to see contemporary society as itself postmodern. 'Postmodernism' can also be conceptualized as a new stage or development within society. Mike Featherstone (1988: 178) describes postmodernism as a 'break from modernity, involving the emergence of a new social totality with its own distinct organizing principle'. Such a principle is not economic progress, improvements in scientific knowledge or more equitable social provision. We have moved on from these priorities, which were attached to the social institutions of industrial capitalism. We now live in a society characterized by 'disorganized capitalism' (Lash and Urry 1987) or more radically, a society of 'simulations' (Baudrillard 1983) where all is image and little, if anything, can be said to be real. Conversely, Marxists such as Fredric Jameson (1984) see 'postmodernism' as a new stage of capitalist relations where the market operates through new information technologies and media to an unparalleled extent.

How might these debates bear on a consideration of the relationship between the environment and society? Put simply, sociologists who stress the importance of 'agency' would argue that what 'exists' depends largely on human actions, thoughts, beliefs and understandings about the world. So what we think of as the 'environment' and how we think we should act towards it, depends on the kind of society we live in and the norms and values it has. This is a social constructionist view, wherein we humans, through our ideologies and practices, fundamentally shape and alter (construct) the world around us. For strong 'social constructionists', social construction is the main influence on social life. For such theorists, we cannot analyse a 'real' objective environment 'out there', but need to consider differing human perceptions of it. The third and fourth sections of this chapter will look in detail at how these approaches theorize environment–society relations. We look at social constructionism as exemplified by Keith Tester (1991), John Hannigan (1995) and Klaus Eder (1996), all of whom are to different degrees attracted to postmodern approaches. We then consider the 'realist' and generally 'modernist' approach exemplified by Peter Dickens (1992, 1996) which conceptualizes the environment as having independent objective properties and powers. For critical realists, there is an environment 'out there', which exists independently of how we think about it, albeit that our conceptualization of the environment influences how we interact with it. Realists tend to stress the relative importance of material (biological, physical, economic) influences on social life. This debate is rather complicated and esoteric, and Michael Bell's (1998: 4) example of how a realist and a constructionist might encounter an elephant for the first time, might clarify things before we become embroiled in the debate (see Box 1.1). Others, such as Bruno Latour (1993, 1999) and Alan Irwin

Box 1.1

Realism and constructionism: a simple illustration

Fundamentally, the realist/constructionist debate is over materialist versus idealist explanations of social life. The tension between materialist and idealist explanations is itself a centuries old philosophical dispute, one that perhaps all philosophical traditions have grappled with in one way or another. There is an ancient fable from India that expresses the tension well.

A group of blind people encounter an elephant for the first time. One grabs the tail and says, 'An elephant is like a snake!' Another grabs a leg and says, 'An elephant is like a tree!' A third grabs an ear and says, 'An elephant is like a big leaf' To the materialist, the fable shows how misinformed all three blind people are, for a sighted person can plainly see how the 'snake', 'tree' and 'big leaf' connect together into an elephant. To the idealist, the fable says that we all have our ideological blindnesses, and that there is no fully sighted person who can see the whole elephant – that we are all blind people wildly grasping at the illusive truth of the world.

Source: Bell (1998: 4)

(2001), have tried to integrate aspects of such approaches in the notion of co-constructionism. In this approach, we construct the environment through our ideas and related practices, but, at the same time, it constructs us. Before thinking about this however, we will consider how environmental sociology has developed historically, which provides part of the explanation as to why current disputes have been so fierce and protracted.

Sociology and the environment

Why is the discipline of sociology relevant to our understanding of the environment? In the minds of many people, including many sociologists, the answer is that sociology should not, and cannot, contribute very much. Sociology is the study of human society and social relations, whereas the environment is often defined as that which is not human, and therefore outside the scope of the discipline. In the popular mind, and as portrayed in the media, the environment is often seen as a 'natural-science issue', because environmental problems include matters such as land, sea and air pollution. However, while the natural sciences have a key role to play in our attempts to address these serious issues, the subject matter of the social sciences is also vital.

Political, economic and social practices and institutions use resources and make various kinds of demands on the environment. As we will see from Chapter 2, many green thinkers blame the ways in which society is currently organized for causing environmental problems. For example, some argue that economic growth, a consumer culture and an urban way of life put excessive strain on natural resources and result in wide-scale pollution. Using the example of 'mad cow disease' in Britain, Box 1.2 introduces some of the ways we might see social institutions, structures and processes as important in understanding the causes and nature of, current environmental 'problems'. An understanding of environment–society relations is as important as scientific knowledge about the technical nature of environmental problems. However, many sociologists have been rather nature-phobic, due to the way in which the discipline has developed historically, as we now consider.

Box 1.2

Mad cows: social, natural or monstrous hybridity?

Bovine Spongiform Encephalopathy (BSE), more commonly known as 'mad cow disease' was diagnosed among beef and dairy herds in Britain in the mid-1980s, becoming an issue of public controversy at the end of the decade and into the 1990s. A key controversy was whether the disease in cattle would be able to 'leap the species barrier' and affect humans with new variant Creutzfeldt-Jakob disease (nvCJD). For much of the 1990s, debates raged as to whether eating beef was safe, whether using medicines containing substances from the spinal cord of cattle was safe, the scale of a potential human epidemic of nvCJD, what caused BSE in the first place, and how best to control the spread of the disease. The case of BSE illustrates a number of points pertinent to a consideration of environment–society relations:

- The *certainty of scientific knowledge* and evidence was questioned, as BSE resulted in a situation where there was scant evidence for the then Ministry of Agriculture, Fisheries and Food (MAFF) to go on.
- *Public perception of environmental risks* or hazards widened to include meat, especially beef, and there was increased concern generally around issues of food safety. As a direct result, beef sales plummeted.
- Periodically during the 'beef crisis' it appeared that *environmental risks were unevenly distributed* in terms of social class for example. At one point, the most 'risky' meat was considered poor quality products such as 'mince', neck and tail meat, and mechanically reclaimed meat (from bones and sinew, or from the head).
- *Government policy was driven by economic considerations* rather than scientific understanding of the problem or public concern. To protect the meat and dairy industries, the government explicitly argued that British beef was 'safe' to eat. More recently in Britain, after the prolonged outbreak of 'foot and mouth' disease in 2001, the MAFF was reorganized and renamed Department of the Environment, Farming and Rural Affairs (DEFRA) as a result of the differently motivated political lobbying by environmental groups, consumer groups and farmers, that the government ministry had acted inappropriately.
- At one stage, some scientific opinion advanced the explanation that BSE-infected cattle had been fed disease-infected feed. This raised a host of questions about *the nature of modern intensive farming practices,* e.g. the feeding of animal-based feed pellets to herbivores such as cattle, the sloppy practices of abattoirs, the impact of breeding and reproductive technologies. The nature of modern British intensive farming problematized the notion of the countryside as 'natural' and undermined the public perception of the rural environment, replacing it with a picture of technical industrialized production.

continued

- Related to this, our **notions of other species** are disrupted also. Is the modern cow, for example, a natural object or a product of corporate biotechnology due to the rigours of scientific breeding techniques to enhance productivity?

So economic, political and social factors shape both our perception of mad cows, and the biological constitutions of the cows. 'British beef' is at the same time an industry, a series of products to be marketed for different kinds of consumer and a symbol of English nationalism.

Differing perspectives in environmental sociology would be likely to have different contributions to make to our understanding of this phenomenon. So for example:

1. **Social constructionists** could be concerned with the ways in which beef came to be perceived as a risk by a significant proportion of the British public. They would be interested in the ways in which pressure groups, government ministries, different groups of scientific researchers and the media went about making claims about BSE.
2. **Co-constructionists** may be concerned about the problems of knowing about BSE and the relationships between expert and lay knowledges. They would also be interested in the overlapping categories and perceptions of the social and the natural.
3. **Critical realists** would probably be concerned with the political economy of meat production in relation to human risks through food consumption. They may also be concerned with animal welfare, and with the political manipulation of scientific knowledge.

The development of environmental sociology: challenging 'nature phobia'

Sociology has a legacy of studying human society as separate from the 'environment' which it tended to define as apart from the social world (Macnaghten and Urry 1995: 203–4). Michael Redclift and Ted Benton (1994: 3) see this as a result of the debates surrounding the development of the discipline in the early twentieth century. In arguing that sociology was distinct from other forms of academic inquiry, early sociologists insisted on the distinctiveness of the 'social' from the 'natural'. This was also a means of countering the influence of biology, which was then often used as a means of explaining social phenomena. Benton (1998) has described this legacy as 'nature-phobia'.

Despite its apparent reticence to examine environment–society relations however, there is much in the legacy of so-called classical (nineteenth-

and early twentieth-century) sociology that may be useful. While not explicitly concerned with the environment, many of the areas of current environmental concern surfaced in the writings of the 'classical sociologists'. As we will see in Chapter 4, Karl Marx, Emile Durkheim, Ferdinand Tönnies, Georg Simmel and Max Weber made inferences to 'nature', and criticized the development of industrial urban society as inhumane, unjust and in some way, 'unnatural'. An explicit attempt to depart from the social preoccupations of the discipline was pioneered in the United States in the 1920s by Robert Park. 'Human ecology' was a development within urban sociology that attempted to incorporate an analysis and observation of the effects of the physical environment on the everyday lives of human beings. It was criticized for underplaying social factors and adopting an uncritical form of social Darwinism in analysing the 'struggle' over land use in the city. Critics of Park and his colleagues however, also suggested it was inappropriate to use material from biology in developing sociological explanations.

It was not really until the 1990s that European sociologists turned their attention to relations between the environment and society. Early accounts still tended to operate with a very firm distinction between the 'social' and the 'natural'. Steven Yearley's (1992) *The Green Case*, for example, tends to use the natural sciences in a fairly uncritical way in order to define what the environmental 'issues' are, and leave the 'sociology' to an assessment of the social consequences of these problems. Howard Newby (1991) called for a reconsideration of sociological attitudes to the environment. Sociologists should be concerned with environmental problems and, in order to study them, they need to question some of their assumptions about the ways in which they do sociology. What Newby hinted at, but did not develop, was that sociologists might have to consider the significance of non-social causes and effects in addition to their usual preoccupation with intra-social phenomena.

In America however, Riley Dunlap and William Catton have since the late 1970s been encouraging sociologists to take the study of the environment seriously. In a famous paper, they made a 'call' for a 'post-exuberant sociology'. This argued for a 'paradigm shift', a fundamental change in approach to the study of sociology. Catton and Dunlap (1980: 34) argued that sociologists operated within a **dominant paradigm**, which was 'human exceptionalist'. It is based on the Western worldview that sees humans as fundamentally different from other species, and holds that social causes shape human behaviour, and that social problems may be overcome by innovations in science, technology and social policy. This human exceptionalist perspective needs to be replaced by a 'new ecological paradigm' which considers the impact of environmental factors on social behaviour, and the impact of social processes on the environment. They suggested that, like other species, we interact with nature and, despite our specific species characteristics, are dependent on the global eco-system.

In reviewing the 'progress-so-far' from the late 1970s until 1993, Dunlap and Catton concluded that, rather than trying to break the mould of sociology, those looking at the environment, tended to adopt a traditional people-centred (or human-exemptionalist) paradigm. This corresponds to what some green thinkers would call an '**anthropocentric**' or human-centred perspective. Sociologists looked at the impact of resource depletion and environmental damage on human health and well-being, lifestyle and values, social stratification, etc., but not at the ways in which beliefs, values and social institutions contribute to environmental damage. Newby (1991: 6) was less harsh in judging environmental sociology at the start of the 1990s. He felt that sociologists were wary of studying the environment due to the political values displayed by many green social theorists. However, sociologists have certainly not shied away from certain subject matter (class, for example) simply because of the political debates surrounding them. Replying to Newby, Stephen Cotgrove (1991) argued that the explanation might be the influence of Marxism in (European) sociology since 1945, which dismissed environmentalism as a middle-class concern. For Cotgrove, increased interest in the environment coincides with the decline of Marxist sociology in the 1980s. Given that some of that interest has been among Marxist and socialist sociologists however, I find this assumption also rather unlikely.

The most probable explanation for sociological nature-phobia is the historical legacy of the discipline and its exclusion of nature as not social. This is partly a reaction against natural-science arguments being used to support politically undesirable explanations of social phenomena, such as the use of evolutionary theory to argue for 'natural' forms of social selection by nineteenth-century socio-biologists (Benton 1991). Benton calls for sociologists to give cautious consideration to biological factors in social explanations, but while sociologists have had much to say about scientific knowledge in the last decade, much of it has not been along the lines he suggested. The amount of work undertaken in environmental sociology in the 1990s, and the lively debates early in this century, would suggest sociologists are no longer as 'nature-phobic' as they were. They are still concerned to draw boundaries between what is social and what is biological however, and most have clung to their traditional paradigm and avoided interdisciplinary analysis. At present, 'nature-phobia' is perhaps best seen as 'science-phobia', and a brief consideration of debates about scientific knowledge seems appropriate here.

The great divide: sociology, 'science' and 'nature'

Given the historical development of the discipline, it is perhaps unsurprising that contemporary sociologists are critical, even dismissive, in approaching the knowledge produced by the natural sciences. Newby (1991: 1) labels environmental science as a discipline characterized by 'technological determinism', and Macnaghten and Urry (1995: 209) argue that scientific theories should always be treated with scepticism due to the cultural assumptions upon which they are based. For John Hannigan (1995: 187), environmental scientists are a key group constructing theories or '**narratives**' about 'problems' connected to the environment. In postmodern accounts in particular, scientific theories about the environment tend to become 'stories' which have social functions, but do not explain anything. This is a form of 'sociological reductionism' which places social analysis in a superior relationship to other forms of knowledge from other disciplines. It ignores the diversity of theoretical positions in the natural sciences and the extent to which some of these have been critically aware of the effects of contemporary political opinions, cultural values and historical events in shaping scientific opinion (see Pepper 1996: 239–95).

Other sociologists have sought to engage with scientific knowledge of environmental issues, and in doing so, have questioned the traditional boundaries of the discipline. Catton and Dunlap's (1978) earliest attempt to draw together social and natural scientific understandings was to propose an 'ecological complex' as a framework sociologists could use to take account of environmental factors in their analysis. Catton and Dunlap see society as composed of a number of aspects: population (e.g. urban dwellers), technology (e.g. cars), culture (e.g. social values, such as consumerism), social structures (e.g. class) and what they term 'personality' (what I would see as psychological factors). The 'environment' is similarly complex with a number of aspects: the natural environment (seas, minerals, etc.), the built environment (e.g. houses and roads), what they call the 'modified environment' (e.g. polluted air, or highly controlled environments such as gardens). Catton and Dunlap argue that all these aspects interrelate with all others, so that, when investigating a sociological issue, one needs to think about all nine 'aspects' that shape human ecology. While this encourages an interdisciplinary approach, Catton and Dunlap give no criteria by which we can assess the relative importance of different processes and, for example, causes of change, or impacts of change, and constructionist critics suggest it takes us too far away from 'the social' (Hannigan 1995: 17).

Perhaps the most important influence on critical realism in Britain is the philosopher Roy Bhaskar. However, critical realist sociologists like Ted Benton and Peter Dickens have questioned Bhaskar's own views about social and scientific knowledges in seeking to respond to Catton and Dunlap's call for a 'new ecological paradigm'. In *The Possibility of Naturalism* (1989), Bhaskar argued that the 'environment' and 'society' could not be studied in the same ways. This is because human societies have distinctive characteristics. Although we can make similar kinds of observations about social and natural phenomena, we cannot analyse them in the same way because the 'structures' that shape social life are so different from those shaping the natural world. Unlike natural structures, social structures are not independent of the activities they govern and do not exist independently of people's perceptions (1989: 30–8). Benton (1985) however argues that these differences are overstated. Social structures can be independent of the activities they govern. For example, states and governments may have legal powers that they never use, and we may not be aware how social structures are shaping our activities and behaviour. We can consider the ways in which social structures and natural structures interact, and Benton (1991) suggests that certain approaches in biology may have similarities to some approaches in sociology and might be drawn upon in order to analyse specific environmental issues.

Peter Dickens (1996: 29) develops such ideas in proposing a 'unified science' that can be used across the boundaries of natural and social science and brings different levels of knowledge into interaction. The theories of life and ideas of natural causal powers of natural objects (such as seas and soils) as understood by biological and physical sciences, can be given additional depth and accuracy when combined with the understandings of contingent social factors which are the prerogative of the social sciences. He has recently argued that in the areas such as health, child development and genetic modification, we need to consider the ways in which genetically determined predispositions, together with prevailing social relations, can explain contemporary patterns of **social**

stratification and human well-being (Dickens 2001: 107). For example, developments in genetic technology indicate that capitalism is modifying human biology to 'make distinct classes of people' (2001: 106). A unified science will not be any kind of absolute truth (1996: 71), but may be a closer approximate understanding of the objects and processes of natural and social life than other forms of knowledge.

Bruno Latour has suggested a different approach to the nature/society divide, which draws on constructionism and critical realism. Latour (1993) argues that if we look at specific environmental issues such as ozone depletion or global warming, social and natural factors interrelate to such a degree that it is meaningless to separate them out. Latour (1987) has proposed 'actor-network theory' (ANT) to explain how the natural and the social world are 'co-constructed'. The scientist, through experimenting on 'nature', modifies and changes both nature itself, and the knowledge we humans have of it. Nature is not 'out there' waiting to be discovered, nor is it simply a phenomenon constructed by human imaginations. Nature is 'co-constructed' by an intermingling of social and natural factors that he calls a 'network'. Whereas Dickens wants to allow for the potential independence of the natural and the social, Latour suggests these categories are inseparable. Although Dickens's account is very different from Latour's, they have similarities in considering the interrelations between the material (resources, technology, biology, economic production) and the ideological (culture, ideas, beliefs, representation). Latour (1993, 1999) can be seen as realist in arguing that humans are not the only actors who construct and change social and natural life; other species also 'act'. Social and 'natural' entities constantly exchange properties within complex sets of relationships, forming **hybrids**.

Social constructionists may think this stretches the boundary of sociology in an improbable and undesirable direction. Interrelations of networks (for Latour) or structures (for Dickens) from nature and society have been seen as 'excessively abstract', and, by undermining our usual ways of thinking sociologically, 'confusing' (Bloor 1999: 97). I concur with Dickens (1996) that strong constructionists go too far in deconstructing scientific claims about the environment. While a naïve realist position would simply endorse scientific claims, critical realism can evaluate such claims with reference to the social context in which they are produced, while also accepting that the natural sciences might have something useful to say about the condition of the environment. The strength of critical realism and ANT is that they question the ways sociologists think, and it is perhaps therefore unsurprising that many sociologists prefer to use social constructionism to look at the environment or not look at environment–society relations at all. The usefulness of social constructionist perspectives on 'science' is that they can demonstrate that 'science' is not 'pure' knowledge. It is a product of the society in which it is developed, and different knowledges may be endorsed by groups with differentially vested interests. I am not convinced that Latour's actor-network theory can be a model for all research on the environment, but it is an interesting attempt to integrate realism and constructionism and challenge the society–nature divide. Human beings do have distinctive qualities as members of actor-networks but, in my view, ANT does not sufficiently account for the ways in which humans are able to exert domination in 'networks' of environment–society relations. We will return to this theme in Chapter 6, looking at the issue of human relations with other species. It is now time, however, to move away from sociology's difficulties with the society/nature divide, to look at what sociologists say about the environment in greater depth.

Social constructionism

Social constructionists argue that things in the world gain their character or 'beingness' from social action, rather than by virtue of any objective properties they may have. This means that our knowledge about the world depends on who we are and where we are. Our knowledge of 'things' does not have any objective basis, but is relative to the culture and the time in which it is produced – it is dependent upon its 'location' in time and space. For Macnaghten and Urry, for example, the environment is conceptualized as a series of cultural constructions with various social functions (see 1998: 44–62). This section focuses on the work of three sociologists who demonstrate differentially strong versions of social constructionism. First we have the strong constructionist position of Keith Tester, followed by the less 'extreme' positions advocated by John Hannigan (1995: 3) and Klaus Eder.

Tester's basic argument is that 'nature' does not have any objective reality. The many objects that make up the 'environment' are products of society, and what we know about reality is a social construct. So the relationship between the 'environment' and 'society' is that the latter produces the former. Martell (1994) argues that Tester conflates his **epistemology** (i.e. his theory of the nature of knowledge, or how we know what we know about the world) with his **ontology** (i.e. his theory about the nature of being in the world, or how things 'are'). Tester does not have an ontology because the 'environment' only exists as a kind of knowledge; it can only be considered in terms of how we know about it. Tester does qualify this extreme position, by saying for example that 'the natural dimensions of being human are fairly trivial' (1991: 32). In saying nature is 'trivial', he is acknowledging there is such a 'thing' as a natural or biological being, it just has insignificant causal powers. Martell is staggered that strong constructionists cannot 'see how big nature is in its ontological powers' (1994: 176). There are a whole host of 'big' environmental powers we are forced to acknowledge, such as the causal powers of the hurricane or tidal wave, or the migratory patterns of birds and various mammals. Tester's position is extreme however, and most social constructionists do acknowledge some sort of 'real' environment.

Hannigan (1995) is concerned with the claims-making activities of environmental social movements and their role in socially constructing environmental problems such as acid rain, loss of biodiversity and the negative implications of biotechnology. Hannigan (1995: 183) argues that the '**postmodern condition**' is taking us into a more environmentally benign '**post-industrial**' future. Like Anthony Giddens (1990, 1991), he sees our current situation as late or high modernity, rather than postmodernity. He uses Ulrich Beck's (1992) well-known and popular thesis of 'risk', which suggests we are now living in a society characterized by 'risk', much of which comes in the form of environmental hazards. There are two aspects to risk. One is the popular perception of risk; the other is actual risk itself. In a 'risk society', for Beck, the political system is not concerned with the production and distribution of material goods, but is becoming preoccupied with the distribution of negative risks. Beck is critical of the social processes of modernization in Western societies, seeing such developments as a key cause of environmental risk. However, he argues that, in contemporary 'high modernity', we are reflexive in our attitudes to modernization, can critically reflect on its drawbacks and respond positively to solve the causes of environmental risks.

My criticism of Beck is that he confuses the social perception of 'risk' with actual environmental risks societies face, so he does not give sufficient weight and attention to the 'reality' of risk. However, Hannigan is more firmly social constructionist than Beck, and rejects the elements of realism in his risk hypothesis. For Hannigan, the environment is best understood in terms of diverse, contestable and culturally/historically relative 'narratives'. He rightly notes that public concern about environmental issues fluctuates over time, and the content of that concern also changes. He suggests that institutions such as the media, environmental pressure groups and the scientific establishment, are key to the social construction of environmental risks such as acid rain, **biotechnology** and loss of **biodiversity**. Hannigan calls his perspective 'social constructionist' because his focus is on the ways environmental issues become seen as problems, but he admits that there is an objective reality to environmental problems and that 'legitimate claims' can be made about the condition of the environment (1995: 30). What differentiates Hannigan's approach from critical realism, is that he argues that, as sociologists, our job is to focus on the process by which claims about the state of the environment are assembled and presented by social groups. So Hannigan adopts social constructionism because it is more appropriate for a sociological approach.

Klaus Eder (1996: 24–32), similarly argues that our ideas about the environment are determined by the culture we live in. We can only analyse nature 'discursively' as a 'symbolically constituted regime'. Hannigan and Tester are concerned with the activities of social movements in creating **discourses** about the environment while Eder looks more broadly at a range of 'symbolic regimes' or 'ways of seeing' environment–society relations. He contends that there are two conflicting discourses about the environment: the modernist 'exploitation discourse' of industrialism, and the 'pollution discourse' of the environmental movement. He argues that, in both modernist/industrialist and environmentalist discourses, nature is regarded as an object, either of utilitarian value, or as requiring some form of human protection. Both these are problematic, and he looks forward to a 'post-environmentalist' future. Eder selects a number of case studies to demonstrate that nature is symbolically constituted: food taboos, different food 'cultures' and the politics of environmentalism. In each case study, Eder over-emphasizes the symbolic aspects of our relationship to the environment however, and there seems little more to the 'environment' than the stories we tell about it.

For both realists and social constructionists, 'nature' is an ideological construct, and popular discourses of nature (ways of perceiving and understanding it), shift in content over time and cultural space. The point where realists and constructionists part company is over the importance of the cultural representation of the environment. I would suggest that social constructionism places too much emphasis on social ideas about nature, and my own sympathies are with realist arguments for the relative autonomy of the environment. Hannigan suggests that social constructionism is perhaps the most fruitful way for *sociologists* to approach the study of environment–society relations, bringing an important sociological approach towards the construction of social problems to bear on environmental issues as hazardous and as a threat to social well-being. Similarly, Tester draws on the sociology of social movements and religious affiliation in his analysis. Burningham and Cooper (1999) give a sober account of the debate, and they suggest that social constructionism is the most appropriately sociological approach to the environment. My own view is that the study of the environment poses particular questions and difficulties for those engaged in the study of society, which social constructionists shy away from. The realists and actor-network theorists are right to argue that the study of the environment means we must change the ways we 'do' sociology.

Critical realism

Realists argue that the environment consists of entities or beings with objective properties, that is, with characteristics that are independent of social processes and human understandings. This means there is a 'real' environment out there, which has certain properties and powers, whether we humans know about or understand those characteristics or not. This is not to suggest that realists are naïve with respect to the social construction of nature. Some have adopted Bhaskar's term 'critical realism' in order to emphasize that one can both accept the idea that the world is composed of 'real objects' with independent properties and causal powers, alongside an understanding of the social construction of that world in different ways by human subjects.

In *Society and Nature: Towards a Green Social Theory*, Dickens (1992) develops a sophisticated critical realist sociological position. He suggests that the relations between any living organism and its environment need to be understood as mutually dependent and mutually constitutive (i.e. we cannot consider the 'beingness' of any one living thing, without considering the environment with which it coexists). Any living being has the potential to act, but the way in which it does so is shaped by the conditions in which the organism finds itself. This is not a one-way relationship but a reciprocal process: the organism can have an impact on its environment, and the environment impacts on the organism. The problem with social constructionism, for Dickens, is that it can only see this relationship as a one-way process, where human thought and action actually constructs the environment.

The key to Dickens's (1992) argument is his realist position that 'things' have independent properties separable from human perceptions of them. So when Dickens talks about 'nature', he sees it as having its own capacities, properties and powers. Human beings continually interact with and affect 'nature' however, and there are 'contingent social circumstances' which shape the environment. Thus, while the environment has natural properties and powers, human practices and social behaviour may shape, enable and constrain the processes of the natural world. For example, the human practices of farming, wildlife conservation and gardening all have implications for the various species of plants and animals, for some are able to flourish and others are 'constrained' from flourishing. Dickens suggests that one of the main benefits of such a 'critical realist approach', is that it can combine a conception of nature as having independent properties and causal powers with the idea that such powers and properties are socially mediated. This provides a compromise between the sociological reductionism of social constructionists, and the sociological *naïveté* of some green social theorists (which we will consider in Chapter 2).

Dickens (1996: 1) has used the example of paid work to look at how social and natural systems intersect, particularly the ways people 'work on nature in order to produce the things they need'. Focusing on the division of labour, Dickens argues, following Marx, that in industrial capitalism people are alienated from nature, nature being transformed into something 'other' than human beings, which humans exploit for its utility and exchange value (1996: 50). So the social and economic relations of modernity have resulted in significant ecological costs as well as the undoubted human benefits (1996: 8). Dickens over-emphasizes the significance of the division of labour at the expense of other aspects of capitalism, and, as Martell (1994: 152) points out, Marx's own discussions of the environment are highly ambivalent and perhaps less useful than Dickens suggests.

Human relations with other species are an undertheorized aspect of Dickens's work, although this has been usefully considered by other critical realists such as Benton (1993), as we will see in Chapter 6. Dickens's (1992) focus is human nature and Marx's concept of 'species being'. **Species being** is Marx's concept of human nature and Dickens develops this, suggesting that our 'nature' is dynamic. Human nature is

shaped by changes in technology and by our working with and on the
environment (see Dickens 1996: Ch. 2). Social constructionists find
the concept of human nature problematically 'essentialist'. **Essentialism**,
in my view, is an overused, imprecise but unfortunately very popular form
of criticism at present. It is used to describe a theory or concept that sees
something as fixed and unchanging. Realists are seen to use fixed
categories and attempt to discover 'truths' about social life, which, given
the social construction of knowledge, is not appropriate. Dickens's view
of species being in humans however is one that changes across time and
space and adapts to different social and economic practices and forms of
organization. We will return to this debate using the example of human
relations with other animal species in Chapter 6, and in discussing
different forms of green thought in the next chapter. Interestingly,
although attracted to postmodern approaches herself, Diana Fuss (1989:
19–21) argues that social constructionist phobia of biological
explanations and of social explanations which try and incorporate natural
science insights, is misplaced. To argue that everything is always socially
constructed is of course, essentialist, because it assumes social
constructions are 'true'!

Burningham and Cooper (1999: 300) argue that the difficulty with the
realist position partly lies in its politics, and they implicitly refer to the
notion of 'essentialism' in their critique. Realists do not simply make the
case that social constructionism is incorrect in failing to acknowledge the
independent causal powers of nature; they suggest that this prevents
sociologists contributing to policy initiatives, as there is no 'truth' to
inform what policy options might be most effective. Burnham and Cooper
argue that to adopt some kind of weaker constructionist argument does
not preclude political action. We can accept the existence of
environmental problems without adopting a realist position, because even
strict constructionists do not argue that there are no 'real' environmental
issues; they merely have a strongly sceptical attitude towards claims about
such issues. So, there are similarities between these seemingly
diametrically opposed schools of thought, and briefly considering
co-construction as a possible compromise seems apposite at this point.

Co-constructionism

Alan Irwin (2001: 162) concludes his text on environmental sociology
with a 'plea for theoretical pluralism and open-mindedness', while

suggesting that 'co-constructionism' might be the most useful theoretical position for environmental sociologists. Irwin places different theorists in different 'camps', than I do here, and sees Yearley and Hannigan as examples of 'cautious constructionists' (2001: 171). Irwin draws upon Latour's (1993) notion of '**hybridity**' in order to make the case that, in lived experience, what is natural and what is social is not easily separated out. The social and the natural are categories which shift and change over time, place and space. In assessing environmental damage, we must bring in the social factors in order to get a clearer picture. Irwin adopts Beck's (1992) notion of a risk society, and argues that living in a risk society forces change in all kinds of social relations, not only relations between the social and the natural. People's notions of the natural are constantly in flux, and sociologists must respond to this by changing the ways in which it conceptualizes 'things', to enable us to see objects as hybrids, that cannot be seen as distinctly natural or social.

The task of sociology is not to produce underlying truths about 'how things are', but to provoke debate, to question, disrupt and challenge (Irwin 2001: 179). Irwin's sympathies for such disruption lie with co-constructionism. While Latour's notion of hybridity is attractive, I am not sure that in the analysis of real research questions, either he, or Irwin actually do see objects, phenomena and processes as amalgams of the social and the natural. The term 'co-constructionism', almost by definition, preserves the separation of the social from the natural (or I would prefer physical). If things are co- or jointly constructed, they are 'made' by two kinds of factors in interrelationship. By allowing the social and the biological to have the potential to be separate, we can account for a wider range of events and phenomena, and examine the complex ways in which physical and social factors interact and interconnect.

Irwin (2001) and Burningham and Cooper (1999, on their own admission), can be read as caricaturing realism, in assuming that realists think they will get at 'the truth' of whatever the matter is. This said, realists, however critical, are not as pluralist as Irwin; they feel critical realism is a way of producing better kinds of imperfect knowledge about the real world. What critics dislike about realism is, they imply, that it is not sociological. I think what is interesting about realism in environmental sociology however, is probably just that – its challenge to the disciplinary boundaries of sociology. In drawing this chapter to a close, we now consider where such boundaries might be drawn, and of what future 'environmental sociologies' might consist.

Table 1.1 Environment and society: competing perspectives

	Environment seen as:	Environmental issues seen as:	Constitution of society	Relations between environment and society
Critical realism	enormously complex and varied series of objects with their own distinct properties and causal powers.	a result of human exploitation and abuse of the environment, in conjunction with known and unknown operation of biological, physical, chemical and other processes. Environmental issues are often interrelated with social issues.	composed of social institutions and social practices. These are sometimes seen as structures which influence/ shape our behaviour and ideas, and as systems of social organization.	nature is mediated through society (especially through science). Social organization has an impact on natural processes. *Key concepts:* expropriation alienation capitalism
Social constructionism	objects about which there are a wide range of differing narratives which shift across time, space and place.	contested; they are selected and promoted by claims making groups such as scientific establishments, environmental pressure groups and the media. Environmental issues are social problems.	emphasize the social construction of knowledge. What we know constitutes a series of competing narratives of social life.	nature is social. Categories of society and nature are constantly reconstituted through one another. *Key concepts:* narratives representation contestation
Co-constructionism	a series of dynamic and changing objects with which human subjects interact. The social and the natural are constructed in relation to one another.	contextual, i.e. we need to understand the social context of environmental concerns. Cultural assumptions influence perception of issues and policy-making. Environmental issues cannot be separated from other social issues.	social institutions and practices and the multiple experiences and representations of social life are mutually constituted.	nature cannot be separated out from the social. *Key concepts:* hybridity boundary-crossing risk

The future of environmental sociology

> Far from standing at the end of a sociological journey, it seems instead
> that we are only at the beginning. (Irwin 2001: 187)

So concludes Irwin's book on environmental sociology, and there are
many possible futures. The different perspectives in environmental
sociology all problematize the nature/society divide and most argue for a
blurring of dichotomous categories. Some sociologists have sought to
address the dramatic economic and social changes of recent decades,
and one response is to argue that all sociologists should reframe their
parameters of study.

In his latest book, which seeks to set an agenda for sociological study in
the twenty-first century, John Urry (2000) tries to unseat perhaps the key
concept of sociology – 'society'. He argues that the sociological study
of different individual societies is no longer appropriate. With the
transnational migrations of people, the exchange of information across the
globe, the increasingly global networks of economic and political activity,
we should move towards the study of what he calls 'global mobilities'.
Urry uses biological analogies to describe these 'flows' of people and
information across the globe which deposit ideas, values, customs and
acquire new ones on their way. Flows operate in networks and come
together in 'scapes' of social phenomena. 'Environmental problems' are
seen here as globalized phenomena over which governments of individual
societies have little control. A more global level of sociological analysis,
moving away from the study of the societies of nation states, and the use
of the notion of 'scapes' might be useful for environmental sociology, and
this is discussed further in Chapter 5. However, I don't think Urry really
challenges the concept of 'society'. Rather, he challenges, and rightly so,
the notion that 'societies' might best be studied in terms of the boundaries
of nation states or distinct regions of the globe. Sociology is still very
much bounded by traditional notions of the social, and this is reflected in
Urry's recommendations for the study of the environment.

In an earlier article on environmental sociology, Urry together with Phil
Macnaghten (1995: 208–16) recommend the constructionist position that
there is no such 'thing' as a natural environment, but a series of
'contested natures' which are constituted and reconstituted over historical
time, geographical and cultural space. They suggest four areas of
sociological study of the environment. First, developing a better
understanding of the social context of scientific knowledge about the

environment in order to better inform policy decisions. Second, the social construction of environmental problems and issues by the media and environmentalist social movement organizations. Third, the social processes that may give rise to environmental problems such as industrialism and consumerism, and **globalization**. Fourth, different cultural constructions of nature: as a collection of non-human species, as a holistic system, as threatened or at risk, etc. and consider environmental attitudes, values and behaviours. They conclude there is no 'simple and sustainable distinction between nature and society. They are 'ineluctably entwined' (1995: 217), and this understanding of the 'environment and society' is a premise of the rest of this book. Indeed, this is so vague an assertion that all environmental sociologists must agree! The areas Macnaghten and Urry suggest are the subject of ongoing and expanding research, some of which has been quite well established. The areas are covered in this book in Chapter 3 (social movements), Chapter 4 (industrialism and urbanization) and Chapter 5 (globalization). The notion of cultural construction can be found throughout Chapters 3 to 6, although none of these chapters is exclusively social constructionist in focus. The areas Macnaghten and Urry set out in their 1995 article can lend themselves to a whole range of empirical research projects reflecting a variety of theoretical positions. So, rather than having a 'sociology of natures' as they suggest, what we will have is a collection of environmental sociologies.

Dickens (2001) suggests another path for environmental sociology, which leads to a blurring of the boundaries between disciplines and a lessening of social science / natural science divide. We have seen his arguments that contemporary developments in genetics, and concerns with the environment and health, compel the future development of systematic links between the social and the natural sciences. He argues that biology, like sociology, is a complex discipline with various theoretical strands, some of which are more compatible with such an interdisciplinary project than with others. Latour's (1993, 1999) actor-network theory, suggests a different agenda again, where we can look at how natural and social entities are linked in certain circumstances, in complex ways, and construct and reconstruct each other.

I am in agreement with Jonathan Murdock (2001: 129) that ANT is mistaken in its ambition to become the overarching perspective to be adopted by those investigating environmental issues. I would say the same about critical realism, although I do not think this is the ambition of all its exponents. I do not think that engaging with natural scientific disciplines is beyond the remit of sociology, but I do think it is one project among many. Murdock argues there is likely to be a continuum of sociological approaches to differing environmental questions and issues, some more conventionally 'sociological', while those who are keen on transgressing disciplinary boundaries will take a more 'ecological' stance. My own preference is for the latter, because in the empirical investigation of environmental questions, insights from a variety of disciplines may be useful. I do not think we are at the point where common methodologies for studying environmental questions or single theoretical perspectives have been developed. In the process of interdisciplinary research, some new common elements of theory and method might emerge. The books in this series are divided between those 'about' society and those 'about' science. The series itself however, can be seen as a project to document the different knowledges multiple disciplines can bring to bear on 'the environment' and as part of the move to encourage interdisciplinary interest.

Conclusion

The sociology of the environment throws debates about the nature of the world and our knowledge of it into particularly sharp relief, because the subject matter forces sociologists to take account of the 'natural'. Sociologists are less 'nature-phobic' than they were a decade ago, although there is a tendency for them still to prefer using established forms of sociological analysis to look at environment–society relations. There have been fierce debates between those using a traditional sociological constructionist position, which sees nature as a series of contested social constructs, and those advocating a realist notion of the environment as multifarious series of complex organisms and entities with their own properties and powers, which are relatively autonomous from how we humans think about them. Actor network theory has sought to combine elements of these approaches in arguing that humans interact with environmental objects and organisms in complex networks, characterized by co-constructionism. When the different varieties of these positions are considered, they may not be so diametrically opposed as it may at first appear, and some forms of weaker social constructionism are actually quite close to some of the more critical of the realist perspectives, and are incorporated in actor-network theory. All these approaches have something to offer those seeking to better understand relations between societies and their environmental contexts.

'Green' theorists and researchers have sometimes suggested that
academia is anthropocentric, and that the social sciences consider
environmental problems from the perspective of human beings and their
conditions of life. I think this is decreasingly so, and there are
developments towards more ecological sociologies. Many sociologists
have not engaged with the theory and research produced by 'the
environmental movement' and have ignored some of the insights green
social and political theory has produced. The next chapter moves away
from sociology to such theory, and looks at what it has to say specifically
about environment–society relations.

Key points

- There are a number of important divisions in sociology between those who
  focus on 'society', social structures and tend to have a more 'macro'
  approach, and those preferring micro-level studies which concentrate on the
  'agency' of human actors. In addition, there are differences between those
  who favour 'modernist' analyses that look for causes of social phenomena,
  and those postmodernists who see social life as too complex for such
  endeavours.

- Sociology has a legacy of studying only 'the social' and seeing the
  environment as outside the scope of the discipline.

- When, historically, sociology has looked at environment–society relations, it
  has tended to examine the social consequences of the human built
  environment, of cities and factories, etc.

- Three important contemporary approaches are social constructionism, critical
  realism and actor-network theory. Despite disputes between proponents of
  different approaches, they are not always as opposed as it might at first appear.

- Social constructionists can have a 'strong' or 'weak' position, depending on
  the extent to which they see the natural world as a product of human
  consciousness and understanding. They emphasize different and competing
  discourses about the environment, and the ways different social groups such
  as the media, 'science' and environmentalists construct environmental
  problems.

- Critical realists see the environment as having objective properties and
  powers. Human society interrelates with nature and changes it; the
  environment in turn affects us.

- Actor network theory argues that people, organisms and natural objects
  interrelate in 'networks' and construct and reconstruct each other.

- In the future, there are likely to be a variety of environmental sociologies using different approaches: some more traditionally sociological and some more interdisciplinary and perhaps more ecological. A host of issues are relevant to sociological study: media representation, social hierarchy and difference and the environment, relations with non-human species, different social relations with the environment over time, space and place, environmental social movement and policy-making.

Further reading

For material on differing approaches within the discipline, I would recommend Derek Layder's comprehensive text *Understanding Social Theory* (London: Sage, 1994). This deals with some complex material very clearly, and the short first chapter in particular gives an outline of the key divisions between approaches. Alternatively, Ian Craib's *Modern Social Theory* (Hemel Hempstead: Harvester Wheatsheaf, 1992) is structured around the differences between structure and agency approaches in sociology, and is both comprehensive and accessible. Useful for making the links between sociological theory and environmental questions, and having detailed coverage of theorists such as Beck and Habermas, is another book in this series, *Environment and Social Theory*, by John Barry (London: Routledge, 1999). See Chapters 4 and 7 in particular.

For an introduction to the area of environmental sociology, an early, but very thorough account is Tim O'Riordan's *Environmentalism* (London: Pion, 2nd edn 1981). More recent is Michael Bell's *An Invitation to Environmental Sociology* (Thousand Oaks, CA/London: Pine Forge/Sage, 1998), the first chapter of which discusses environment–society relations in a very introductory but quite interesting way. Steven Yearley's *The Green Case: A Sociology of Environmental Issues, Arguments and Politics* (London: Routledge, 1992) is particularly useful for material on the environment and the sociology of development. Alan Irwin's recent *Sociology and the Environment* (Cambridge: Polity, 2001) concentrates on particular issues and themes such as 'risk', policy-making, and science and technology. The lengthy introduction provides an excellent overview of some of the debates discussed in this chapter. The best text, in my view, is Luke Martell's *Ecology and Society* (Cambridge: Polity, 1994), which is strong on theory, detailed but very accessible, and agues a case in addition to reviewing current debates.

On the differing approaches in sociology, see Peter Dickens *Society and Nature* (Hemel Hempstead: Harvester Wheatsheaf, 1992) on critical realism, and John Hannigan *Environmental Sociology* (London: Routledge, 1995) on social constructionism. On ANT, see Bruno Latour's *We Have Never Been Modern* (Hemel Hempstead: Harvester Wheatsheaf, 1993). Useful articles from the journal *Sociology* include: Phil Macnaghten and John Urry (1995) 'Towards a

Sociology of Nature' (vol. 29, 2: 203–20), Kate Burningham and Geoff Cooper (1999) 'Being Constructive: Social Constructionism and the Environment' (vol. 33, 2: 297–316), Peter Dickens (2001) 'Linking the Social and the Natural Sciences: Is Capital Modifying Human Biology in its Own Image?' (vol. 35, 1: 93–110) and Jonathan Murdock (2001) 'Ecologising Sociology: Actor-Network Theory, Co-construction and the Problem of Human Exemptionalism' (vol. 35, 1: 111–33). A good summary of the historical development of environmental sociology in the United States is provided by William Freudenberg and Robert Gramling (1989) 'The Emergence of Environmental Sociology: Contributions of Riley E. Dunlap and William R. Catton Jr', *Sociological Inquiry*, 59, 4: 439–52. In addition, Catton and Dunlap's (1980) call for a new framework for the sociological study of the environment, 'A New Ecological Paradigm for a Post-Exuberant Sociology', *American Behavioral Scientist*, 24, 1: 15–47, might also be of interest.

Discussion questions

1 Think of a recent environmental issue. What claims have scientists made about it? What have environmental activists said about it? How has the media represented it? How accurate do you think your own understanding of the issue is?

2 If most sociologists agree that there 'is' an environment 'out there', and that our understanding of it is socially constructed, why are they squabbling?

3 What would be an example of a 'hybrid' object?

2 Environment and society in green social and political theory

This chapter will:

- outline some general characteristics of 'green' political and social thought
- critically discuss the theories of 'anthropocentrism' developed by 'deep' ecologists
- consider how social ecology has attempted to combine an analysis of intra-human oppression with the idea of environmental exploitation
- compare such attempts with eco-socialist analysis of the role of capitalism in degrading the environment
- compare and contrast the different ways eco-feminism has theorized relations between social exclusion, particularly gender, and 'nature'

Introduction

This chapter, like the first, is primarily theoretical, but it looks at the environment–society relationship from a 'green' or environmentalist perspective, rather than from the disciplinary perspective of sociology. Some of the material on 'green' political and social thought has tended to emphasize the best known of the green positions, deep ecology, and concentrate on an exposition of its themes (Dobson 1990; Goodin 1992). Other authors have identified a number of strands of green political and social thought. For example, David Pepper (1996) and Carolyn Merchant (1992) both distinguish 'deep ecology' and 'social ecology', and have a third category of what Pepper refers to as the 'New Age Tendency' and Merchant as 'Spiritual Ecology'. Others, such as Robin Eckersley (1992) have looked at green theory from the perspective of established political thought, examining the green credentials of different political theories.

The focus in this chapter is a critical consideration of four key strands of contemporary green theory: deep ecology, social ecology, eco-socialism and eco-feminism. I will consider the 'spiritual' elements of green thought in the sections on deep and feminist ecology, and the relations between ecology and other radical thought, in those on eco-feminism, social ecology and eco-socialism. Sociologists have been particularly concerned with issues of social inclusion and exclusion based on class, gender, 'race' and/or ethnic hierarchy, and on relations between 'developed' and less or 'underdeveloped' societies. The chapter considers how green social theorists have understood questions of social difference and inequality, and the extent to which they consider these relevant to the relationship between the natural environment and society. Before reviewing the diversity of green social and political thought however, it is necessary to begin by defining what 'green' perspectives have in common.

'Green' theoretical approaches

Andrew Dobson (1990: 13–23) sees the division between environmentalism and ecologism as a key dichotomy in green political thought. Some green activists and thinkers such as Jonathon Porritt (1986: 5) have dismissively referred to environmentalism as 'light green'. Arne Naess (1973) implies that it is superficial by calling it, 'shallow' (as opposed to his own 'deep' ecological position). Environmentalism is usually defined by its concern with conservation through **technocratic** means (Benton 1994: 31), and 'environmentalists' are often represented in the literature as reformers who do not consider radical change in political, social and economic institutions and processes necessary. For example, some have argued that currently existing social practices such as those associated with consumer capitalism, can be modified in order to preserve the environment (Elkington and Burke 1987).

For Merchant (1992: 1), it is a sense of environment–society relations as in some sort of crisis, which defines 'ecological' perspectives. She calls such perspectives 'radical ecology', but I think Dobson (1990: 13–18) is right to suggest that ecological thought is by definition, radical. Merchant argues that ecological thought sees environment–society relations in terms of the human domination of nature, with which I would agree. However, as we will see throughout this chapter, different schools of ecological thought have different analyses of exactly how and why humans can be said to dominate nature. They also have different understandings of the ways in which social differences affect environment–society relations. Ecological perspectives concur that current forms of social and economic organization constitute a threat to the long-term well-being of the planet and must undergo radical change. However, the political means accepted by green thinkers and activists does not fit neatly into the category of 'radical' political action, but also embraces reformism. The distinction between environmentalism and ecologism may not be hard and fast. As we will see in Chapter 3, the political philosophy adhered to by various green organizations does not necessarily tie in neatly with a certain form of political action, and many groups endorse both radical and reformist political strategies.

This chapter discusses four strands within ecological or green theory: social ecology, deep ecology, eco-socialism and eco-feminism. The best known is deep ecology, which sees '**anthropocentrism**' or human-centredness as the key cause of environmental problems and as the structuring principle of contemporary society. Social ecology analyses ecological problems and environment–society relations in terms of 'intra-human' social hierarchies of gender, class, 'race', and the political formations of state and nation. Eco-socialists generally agree that the economic structures and social formations of capitalism shape environment–society relations. Eco-feminism sees the social domination of women and the social domination of 'nature' as sharing certain common elements, and looks also at the ways different forms of social hierarchy structure our relationships to the natural world.

Deep ecology: an anthropocentric society?

Deep ecology directly problematizes the relationship between environment and society. It conceptualizes the human species as in a relationship of domination over nature, and suggests that human exploitation of the environment is currently precipitating a crisis in environment–society relations. We have damaged the environment to such a catastrophic extent that radical and drastic measures are necessary to halt such destruction, and this requires that we change the way we conceptualize nature.

Deep ecologists argue that we must move from human-centredness or 'anthropocentrism' as the key structuring principle of social organization, to a nature-orientated '**biocentric** or ecocentric' (Naess 1990: 135) way of thinking. The most significant contribution of deep ecological thought has been the concept of 'anthropocentrism', which we will begin by considering. We then turn to some of the problems with deep ecological thinking: first, the ethical difficulties encountered in making a case that the environment has value, second, the anti-humanism which runs through some deep green perspectives. Finally, we look at debates around modernism/postmodernism in relation to deep green spirituality and 'affinity' theory.

The human domination of nature

Deep ecologists have tended to argue that modern Western societies are 'anthropocentric'. According to Dobson (1990: 63) this can be defined both as 'human-centred' and 'human-instrumental'. An anthropocentric society is one with a dominant worldview in which the non-human natural world is both conceptualized and treated in terms of means to human ends. In part, this worldview is a product of the European Enlightenment which placed human beings and their faculties (especially that of reason) in a pre-eminent position with respect to the environment.

While many deep ecologists acknowledge that the exploitation and domination of the planet is linked to social forms of domination and exploitation among human beings, they usually see anthropocentrism as the most deep-seated form of domination. Warwick Fox (1989) asserts that human domination of the environment accounts for other forms of social domination, such as those based on class, race and gender. This position is reductionist. It does not account for social complexity in positing that one theory of **social exclusion** can account for a whole range of differing forms of social domination. For Fox, anthropocentrism is the a priori form of **social domination**. The latter preceded all forms of intra-human domination, and is largely irrelevant to any discussion of the human social domination of nature. Developing new concepts with which social theorists can analyse environment–society relations is undoubtedly important, and argument that the social domination of nature requires separate consideration from the differing forms of social domination that structure human society, is also an important contribution. However, in making the case for separate consideration, deep ecologists tend to assert that anthropocentrism should be analysed in isolation from intra-human dominations such as those based on class, 'race' and gender. This moves green theory towards an anti-humanist position, where humans as a species are seen to be collectively responsible for environmental destruction, rather than having differing forms of social organization, worldview and values, as well as differing forms and degrees of environmental impact.

In thinking sociologically about deep green thought it is important to distinguish between ontological and epistemological dilemmas. Ontologically, deep ecologists may argue that contemporary Western societies are premised on a worldview that objectifies nature. They may contend that certain social institutions and practices result in environmental damage, and undertake empirical research in order to provide some evidence for their assumptions. However, deep ecologists have run into difficulties with the epistemological basis of the theory of anthropocentrism, particularly as they have attempted to argue from an ethical standpoint. Some argue that we should care for the environment because we will benefit (or, more pointedly, because if we do not, we will endanger our own species). For Bill Devall (1990: 70), human-centred environmental preservation ethics and nature-centred preservation ethics are not opposing positions, for all forms of nature conservation can be seen as human 'self defense'. There are instances in which human need and the needs of other species or eco-systems may come into conflict however. For Naess, what is best for the Earth as a totality should predominate in any ethical quandary, but in practical application, the implications of this are not always readily apparent. It is unclear to me for example, why fleas and their 'hosts' such as the domestic dog or cat, should have an equal right to live and blossom.

Some have seen such human-instrumental reasons for environmental ethics as themselves anthropocentric and advocate instead an ethic of **'intrinsic value'**, according to which objects are seen to have value in themselves rather than having value in terms of their functions for other things. For Naess 'every living being has intrinsic value' (1990: 135), and he argues for a position of biocentric egalitarianism, that is, that all living beings, including humans, have equal intrinsic worth, because all species are equal in their importance to the planet as a whole. Robin Eckersley (1992) argues for an 'ecocentric ethics' which affirms 'our embeddedness in ecological relationships' (1992: 53) and will allow '*all* beings (not just humans) to unfold in their own ways' (1992: 172, original emphasis). However, she does not discuss differences among and between 'all beings' which may necessitate different ethical treatment, or how this might impact on human society. For Mary Mellor (1992: 85), such biocentric egalitarianism is almost inevitably anti-humanist.

Luke Martell (1994: 80) rightly argues that we need different sorts of value, and different criteria for attributing value, for different environmental phenomena. For example, he suggests that humans and (certain higher) animals share a capacity to gain a sense of well-being from their experiences. On the basis of **sentiency** (ability to experience pleasure and pain), they should be attributed intrinsic value. Plants, rocks or streams do not have such capacities, and so the value we attribute to them is not intrinsic (value in and of itself) but extrinsic (deriving from their objective properties, but attributed because of their value for something else). Deep ecologists have floundered in their search for a philosophical basis for intrinsic value because they will not ethically differentiate like this in order to account for biological diversity and environmental complexity. Instead, they have tended to fall back on a spiritual argument for intrinsic worth in 'all' nature. The notion of 'transpersonal ecology' developed by Warwick Fox (1989) describes a process where the environment is seen as acting both on and through human beings. When humans acknowledge this process they experience identification with and compassion towards nature.

Val Plumwood (1993: 17) critiques such approaches as stressing an individualistic and psychological 'affinity' with nature, where we are encouraged to develop an 'ecological consciousness' that personally 'attaches' us to the natural world. We are given no guide as to how such an ecological consciousness might be acquired, or how this will enable concrete change of social, political and economic institutions and practices. David Pepper (1996: 115) is not so much critical of the individualism of deep ecological thought, but of its anti-humanism and mysticism. I think deep ecologists such as Fox (1990) have qualified elements of anti-humanism present in some of the works of the 1980s, and I do not share the hostility of some socialist and social ecologists to the spiritual elements of some deep ecology and eco-feminism. However, in emphasizing the absorption of humans in 'nature', deep ecologists fail to account for relations of both similarity and difference between humans and other parts of 'nature'. An acknowledgement of such difference, and a conceptualization of human–nature relations which positively revalues the natural world in all its diversity, requires more than simply dissolving the boundaries between humans and 'nature'.

Modernity, postmodernity and deep ecology

Some critics of deep ecology have been arguing from a 'modernist' perspective that deep ecology does not sufficiently account for the complexity of social life. One of the first feminist critics of deep ecology was Australian socialist and eco-feminist Ariel Salleh, in an article published in 1984 – 'Deeper than Deep Ecology'. She contended that deep ecology ignored women's lives and experiences. It spends too much time constructing an abstract environmentalist ethics, and ignores the material relation of women and men to both production and reproduction. More recently, Salleh (1997) has argued that deep ecology makes social difference and division invisible, when what is necessary is the theoretical and practical political integration of insights from ecology, feminism, socialism, anti-militarism and postcolonialism. Sharon Doubiago (1989) has also argued that there is entrenched sexism in the idealization of human–nature relations within much of deep ecological thought, which stresses the importance of 'wilderness' preservation, and of our experience of **wilderness** in walking, running, cycling, even hunting in it. This positive experience of 'nature' is not for the infirm, elderly or, in many cases, for women, but is a young, male experience idealized by white deep ecologists. For Salleh, the problem is that deep ecologists suggest all human societies are equally to blame, however socially disempowered they might be, or, as Mellor (1992) has pointed out, however light the burdens many societies in the Southern Hemisphere place upon the globe. Fox (1989) replied to such criticisms by asserting that deep ecologists have been misunderstood, and do not presume all humans are equally responsible. However, deep ecologists do not discuss the responsibilities of international capitalism or imperialism, gender or class hierarchy, etc. for environmental exploitation, as this would be human-centred.

Deep ecological conceptions of human-centredness reflect elements of postmodernity in their trenchant critiques of the social and economic structures of modernity, and of the theoretical paradigms (such as scientific rationality) that characterized the development of modern societies. Some have seen the political worldviews of modernity as essentially the same anthropocentric positions – part of what postmodern theorists would term the Enlightenment 'project' of modernity.

However, other elements of deep green thought are thoroughly antithetical to postmodern ways of thinking. While critical of the mechanistic scientific theories which have been a legacy of the European 'Enlightenment' (see Eastlea 1981), some deep ecologists have 'adopted' certain forms of what they see as 'holistic' scientific inquiry. Deep ecologists have drawn upon certain scientific perspectives to vindicate their view of the world as an interrelated web of interdependent systems. Activist and theoretical physicist Fritjof Capra (1983: 77) has deployed egalitarian particle physics, the deep greens' favourite scientist, James

Lovelock (1979, 1989), has used systems analysis in developing his well known 'Gaia hypothesis', which argues that the planet is kept healthy by mutual interdependence with the organisms which live upon it. Deep ecologists do not consider that 'nature' or the environment is socially constructed, although they do argue that the dominant Western conceptions of 'nature' significantly shape how we treat the environment. Deep ecologists base much of their theorizing on an 'essential' nature, which can best be seen in the 'wilderness' and which humans must refrain from 'contaminating'. Critics see an anti-humanism running through such theories in both the social and natural sciences (Pepper 1993: 115). This homogenization of human society as environmentally destructive is the key criticism which can be levelled at deep ecology, and the sensitivity to differences within human society and their impact on the environment is the key strength of the other ecological approaches.

Social ecology: social domination and environmental exploitation

For social ecologists, environmental abuse and exploitation is the direct result of the domination of some groups of human beings by other groups of human beings, or as social ecologists often describe it 'intra-human domination'. Social ecology draws upon other traditions in social and political thought, mainly left-wing anarchism, for an analysis of ecological problems in terms of human social hierarchy based mainly around class, gender and 'race'.

Social domination and nature

The American writer and activist Murray Bookchin, is usually seen as the 'founder' of social ecology (Roszack 1989), and has also been described as the most significant anarchist thinker of contemporary times (Marshall 1993). For Bookchin, the exploitation of humans by other humans is the key to explaining the human exploitation of the natural environment:

> the very concept of dominating nature stems from the domination of human by human, indeed, of women by men, of the young by their elders, of one ethnic group by another, of society by the state, of the individual by bureaucracy, as well as of one economic class by another or a colonized people by a colonial power.
>
> (Bookchin 1980: 62)

One of Bookchin's most important contributions to green social thought is the idea that all humans are not equally responsible for environmental destruction. Bookchin has been hostile to deep ecological theories of 'anthropocentrism', which see human domination of the environment the primary form of social domination and one for which all humans are collectively responsible. He argues that the domination of nature came after the domination of human by human, is of secondary importance to intra-human domination and is caused by it (Bookchin 1990: 44). In his most substantial work, *The Ecology of Freedom* (1991), he gives a complex account of how social hierarchies emerged with the oppression of women through 'patricentricity' ('father-centred' society), proceeding to examine the exploitation and oppression of other groups of humans, socially stratified according to 'race', class and sexuality. Bookchin argues these oppressions adopt different forms (i.e. they are imposed by different social institutions and practices) and degrees of severity across different cultures and throughout historical time. The domination of the environment, for Bookchin, is a direct consequence of these forms of human domination (1980: 63). Particularly important was the emergence of the modern state, which reinforces all social hierarchies, and has a vested interest in depleting natural resources (Bookchin 1971: 27).

In his analysis of the impact of these different kinds of exploitation/oppression on environment–society relations, Bookchin provides a complex account that avoids the anti-humanism of some deep ecology. Mary Mellor (1992) sees Bookchin as pointing green social theory in the 'right direction', as social ecology is clearly aware of the social causes and consequences of environmental destruction and understands that:

> Human society is fractured by race, class and sex. It is not right to
> hold the poor or people of the South responsible for developments by
> the rich white North. Women cannot be held responsible for the
> actions of men. The ecological unsustainability created by the
> grabbing and exploitation of land by dominant groups in dominant
> nations must not be paid off by population control imposed upon the
> poor and landless.
>
> (Mellor 1992: 106)

In the early 1990s, there was a mutual antipathy between social and deep ecologists, as represented by the 'dialogue' between Bookchin and fellow American eco-activist and deep ecologist, Dave Foreman (see Chase 1991). Bookchin has been highly critical of deep ecology as a form of mystical 'wilderness reverence' (1991: xviii). Whereas deep ecologists have a tendency to stress the similarities between humans and certain higher mammals, and to deconstruct the divisions between species, Bookchin emphasizes the distinctive qualities of human beings (1991: xix). Usually, this means that he sees human difference in terms of human superiority. Because human beings are social and animals are only 'communal', and because only humans possess 'reason' and are able to reflect on their actions towards the environment, we constitute 'second nature' above all other animal and plant species which are 'first nature' (1991: xxi). Bookchin over-emphasizes relations between differentially stratified humans at the expense of a more thorough analysis of environment–society relations. He homogenizes 'the environment' rather accounting for the different ways in which, and degrees to which, various aspects of 'nature' are subject to domination.

A modernist ecology? Science, technology and rationalism

Deep ecologists and eco-feminists have criticized 'Western science' and the culture of rationalism associated with post-Enlightenment Western society. Social ecologists however, do not deconstruct scientific knowledge and the related rationalist paradigm of Western modernity. Bookchin concedes some of the criticisms made of scientific rationality, but feels that much of this is due to the assimilation of science by the 'established social order' (Bookchin 1971: 57). Unlike deep and feminist ecologists, social ecologists do not see scientific epistemology as part of the problem, but tend to assume that 'science' is potentially neutral.

In his early writings, Bookchin discusses scientific developments in relation to technological change. He argued that technological developments have placed the West in a potentially revolutionary position (Bookchin 1971: 33) where we are 'on the threshold of a post-scarcity society' (1971: 10). This means potential human liberation from both 'want' and 'work' through the application of new technologies. Problematically, Bookchin fails to differentiate between the range of ontological and epistemological positions in the natural sciences, some of which, as eco-feminists have argued, may be vehicles of the kinds of social domination to which he is opposed.

Bookchin is keen to advocate the usefulness of the ecological sciences in the development of radical political arguments. As Bookchin understands it, ecology finds no evidence of hierarchy in nature; rather, it posits webs of interdependent relations between natural phenomena and the eco-systems in which they are embedded (Bookchin 1971: 58–60). This is similar to the arguments of the nineteenth-century Russian anarchist Peter Kropotkin. In *Mutual Aid* (1955), Kropotkin argued against the social Darwinism of his time, that the logic of natural evolution is not competition within and between species in order that the 'fittest' survive, genetically prosper and reproduce. Similarly, for Bookchin, nature is unified despite its diversity, and species exist in relations of mutual interdependence and cooperation (1991: 26). Bookchin considers social hierarchy to be the enemy of the 'natural order'. As an anarchist, he sees this as imposed upon human society, and, like Kropotkin, thinks it is falsely attributed to the natural environment.

The Ecology of Freedom (first published in 1982, second edn 1991) outlines an evolutionary model of human social development. Bookchin suggests that social hierarchy emerged in the early Neolithic period with the establishment of rudimentary forms of government and the development of warrior groups to protect and extend territory. There are similarities between Bookchin's account and some eco-feminist approaches, for he argues that such developments were also associated with a shift in social organization from a matri-centric (mother/female-centred) principle to a patri-centric (father/male-centred) one. Forms and ideologies of social domination have clouded human vision so we cannot see the correct path for human development in symbiosis with the natural environment. Bookchin's solution is for us to rationalize our way back (1991: 254–6). *The Ecology of Freedom* is extremely ambitious and, in attempting such broad coverage, it leaves much unexplained. Perhaps the strongest criticism is that Bookchin does not attribute agency to nature, the natural world is not seen as having properties and powers in and of itself (to affect other natural phenomena for example by one species killing another for food). Strangely, Bookchin does not say much about human domination and exploitation of the environment, something which has been key to deep and feminist ecological accounts.

In contrast to deep ecologists, Bookchin adopts a hierarchical approach to environmental ethics. Humans are 'second nature' (defined by having 'culture'), and, as thinking and reflexive beings, they can act as the 'voice' of first nature – the **eco-system** (1990: 182). He is dismissive of deep ecology's case for an ethics of intrinsic value, but argues that the environment only has 'rights' if, when and how, human beings decide to confer them on 'it' (1991: xxxv). Humans, however, do have intrinsic value by virtue of their powers of reason and creativity. He argues for an ethical position that understands forms of hierarchy in nature which do not imply domination. He talks of the different levels of sentience between animals, for example, and the relative importance of different organisms in certain biotic environments. Bookchin sees human society as materially constrained by 'first nature' in terms of resources, but these constraints are left under-theorized. Unlike deep ecologists, who stress the links between certain species of higher animals and humans, and conceptualize humans as very much a part of nature, Bookchin sees the ethical divide between humans and the 'environment' as vast and unbridgeable. He has a conservationist notion of humanity as 'stewards' of nature, underestimating the potential conflict between human interests and the multivariate forms of interest that may be present in nature (Plumwood 1993: 16). In addition, Bookchin cannot use the concepts of **exploitation**, **oppression** and domination in application to the environment in the same ways he does to humans, and would see any attempt to do so as an 'anthropomorphic projection' (1991: xxiii). In some instances, deep ecological critiques of Bookchin as anthropocentric are rightly made, for he takes us so far in a humanist direction – he suggests that the environment requires human control and intervention in order to achieve its full potential (1989: 203).

While Bookchin's social ecology is an important critique of the stronger and anti-humanist forms of deep ecology, he goes too far in stressing the social causes and consequences of environmental degradation. The key point of deep ecology is that the environment–society relation has been a forgotten aspect of social theory and research, and that such a relation is a form of domination with devastating consequences, not only for humans, but, most immediately, for plants, the diversity of non-human animal life, waterways, forests, etc. Bookchin's social ecology can provide an important link between analyses of human–environment relations, and critical perspectives on intra-human relations such as socialism and feminism. In the wide scope of his coverage of intra-human oppressions, Bookchin does not analyse each in sufficient detail, nor elucidate exactly how any one form of domination results in certain forms of environment–society relations. 'Bookchinite' social ecologists have often been hostile to other positions. Although Janet Biehl (1988) once attempted to develop connections between social ecology and feminist analysis, by 1991 she had dismissed all kinds of eco-feminism. This is a pity, because some eco-feminist and eco-socialist analyses could be usefully deployed in developing Bookchin's ideas on the interconnecting forms of social hierarchy in relation to the environment.

Val Plumwood (1993: 15) criticizes social ecology's refusal to question the Western conception of rationality, making the point that the elevation of human reason in modernist social thought has been a key justification for human superiority over the natural world. Other social ecologists, such as Theodore Roszak (1992), have roused Bookchin's ire by trying to develop approaches that question and reformulate rationalism. Bookchin would see Roszak's 'ecopsychology' as individualist, but Roszack's aim is to combine the rationalism of social ecology with deep ecology's politically motivating understanding of humanity as part of, and continuous with, the environment. My own view is that the relationship between specific forms of social domination and the domination of the environment, has been more successfully achieved by Marxist and socialist ecologists with respect to the impact of capitalism, and eco-feminists with respect to gender relations, than it has by social ecologists. As will be argued later, eco-feminism also provides a balance between the ecological reductionism of deep ecology, and the sociological reductionism of Bookchin.

Eco-socialism: capitalism and the commodification of nature

Historically, socialism has had little to say about the environment, and when it has, it has shown little sympathy toward environmentalist perspectives. Since the early 1980s however, a number of socialists and Marxists have sought to explore the relationship between 'red' and 'green' thought and concerns (e.g. Bahro 1984; Weston 1986; O'Connor 1989; Pepper 1993; Benton 1996). Like eco-feminists and social ecologists, eco-socialists are interested in the relationship between **social inequality and difference** among humans, and the domination of 'nature'. Critics of eco-socialism have argued that it accepts the premise of the necessity for an industrialized society aiming at achieving economic growth and expanding consumption of resources (Porritt and Winner 1988: 256).

Some eco-socialists in the 1980s responded to such critiques by defending the industrial production economy. They tended to argue that it is the specific use that capitalism makes of industry, not industrial production per se, which constitutes the problem. Such use is directed, according to Joe Weston, towards the creation of profit rather than the satisfaction of social and economic 'need', and the consequence, poverty, is the main cause of environmental problems (Weston 1986: 4). Weston's solution to environmental problems is a radical redistribution of wealth both within the nation states of the developed world, and across the divide between the affluent West and the not so affluent 'rest' of the regions of the globe (1986: 156). It is not clear however, what is specifically ecological about such theorizing. Weston does not critically evaluate concepts of wealth and of need, and leaves unanswered the questions raised by deep ecologists about the unsustainable nature of affluent, consumption-driven, Western societies.

Other eco-socialists, such as Michael Ryle (1988), focus on 'communitarian' or so-called utopian decentralist socialism, as exemplified by William Morris, G.D.H. Cole and Robert Owen, and contend that this has much in common with certain aspects of green social and political thought. For Ryle (1988: 20) socialism is a necessary grounding for ecological theory, for it can provide ecological theories with an understanding of social and political life that cannot be analysed in the same ways as human–environment relations (1988: 7). Dobson (1990), however, contends that much of the social and political programme of green theory can be seen to be anarcho-communist not socialist. While left-wing anarchism is a clear strand of influence in green social thought, this has had an often acrimonious relationship with developing eco-socialist approaches. Since the early 1980s, some eco-socialists have utilized some of the concepts of classical Marxism in order to understand environment–society relations, sometimes linking this project to elements of green anarchism in social and deep ecology (Pepper 1993), sometimes not (Dickens 1996). Dickens (1996), Pepper (1993) and Benton (1996) have sought to demonstrate that elements of 'classical Marxism', that is, the ideas of Marx and Engels, are potentially useful in developing a 'green' social theory.

The legacy of Marxism

Peter Dickens (1992) has argued that Marx has an important contribution to make to the understanding of environment–society relations. Marx has, in Dickens view, a **dialectical** conception of relations between society and the natural environment. Humans depend on the natural environment for their existence, and are shaped by and in turn alter, their surroundings. In addition, we are dependent on the natural world for the realization of our intellectual and aesthetic powers. For Dickens (1996), it is the social

organization of labour power that is the key influence on environment–
society relations. Of necessity, we work on nature to produce the things
we need. Certain ways of organizing labour in capitalist societies, around
the production of goods for the market, on the basis of increased
production and consumption to satisfy the profit motive, means that the
natural environment is exploited. Marx's notion of **alienation** is usually
deployed to analyse people's disaffection with their paid employment, but,
according to Dickens, it can also explain our 'alienation' from nature,
which is transformed into objects valued only as property or commodities
under capitalist relations. For socialist eco-feminists such as Ariel Salleh
(1997), however, an important omission in eco-socialist theories of labour
is the absence of gender. I would agree that eco-socialists need to take
analyses of gender difference and inequality more seriously, and relate
production to reproduction, to avoid marginalizing women's 'caring'
work.

Ted Benton (1996) has emphasized the concept of 'species-being' in
the early writings of Marx. This is really a notion of what it means to be
human, and, according to Marx:

> Species-life, both for man and for animals, consists physically in the
> fact that man, like animals lives from inorganic nature. . . . Man lives
> from nature, i.e. nature is his body, and he must maintain a continuing
> dialogue with it if he is not to die.
>
> (Marx 1975: 327)

The concept of species-being brings us to the interface between the social
and the biological, which, as we saw in Chapter 1, has been something of
which the vast majority of sociologists have been wary. An examination
of what it means to be a human or other kind of animal is important
however, and I would agree with Salleh (1997), who argues that both
socialism and feminism need to take account of the material embodiment
of us human animals. Benton's (1989, 1993) eco-socialism is developed
around these key themes in Marx of natural limits and human nature
(species-being). As we will see in Chapter 6, he has made a powerful case
for a reassessment of human–animal relations, taking into account the
diversity and specificity of non-human animals' 'species-being'.

In developing an eco-socialist perspective, Pepper (1993) outlines various
points of possible cohesion between Marxian socialism and green
anarchism. He argues the Marxian dialectic (where social relations are
based on conflict, alternative positions and social movements develop in
contestation to the status quo, and result in a new synthesis of social

relations) can be applied fruitfully to environment–society relations. We humans materially interact with nature and thereby change it (1993: 111–13), and this is both material (physical change to the environment, forms of human labour power and technological development) and ideological, that is, it influences how we think about nature. We are in a dynamic dialectical relationship with the environment, changing nature as it in turn is changing us. As economic globalization means industrial manufacture moves to the poorer countries of the world, where labour is cheaper and there are fewer pollution controls, the links between industrial production, environmental damage and exploitation of workers are perhaps stronger now than when industrial manufacturing was at its peak in affluent locations.

Critics (e.g. Porritt 1986) have argued that eco-socialists ignore the degradation of the natural environment under state socialism in Eastern Europe for much of the twentieth century. I think eco-socialists do take this on board (e.g. Redclift 1987: 45; Gare 1996) but they argue that environmental exploitation is not systemic within socialism as a form of social and economic organization. James O'Connor (1989) contends that, whereas capitalism as a system of economic production premised on the profit motive is inevitably polluting and depletes natural resources, socialism, based on the principles of reciprocity and redistribution, does not have a built-in propensity to pollute. Pepper argues that environmental risks are disproportionately visited on poorer populations of both rich and poor countries. This fleshes out Beck's (1992) risk society thesis, where the environmental risks of capitalist production are unequally distributed according to wealth. Green socialism will emerge, Pepper argues, as ecological improvement is predicated on the undermining of capitalism as a system of production. An eco-socialist mode of production will produce and distribute goods and resources according to need, not the demands of capitalist consumerism (1993: 146). O'Connor and Pepper assume that, given the limits on 'need', a socialist system of production in contemporary society would place less pressure on natural resources and be better placed to operate within natural limits (see Benton 1989, 1996).

For Pepper, Dickens and Benton, the key point is that capitalism as a system of economic production and social relations, alienates us from the natural environment and also from ourselves. Capitalism encourages us to see the natural world as a series of commodities for use rather than things we work with in reciprocal relationships.

Social inequality, the degraded environment and class politics

For eco-socialists, basic socialist principles are also ecological ones. Eco-socialists have been concerned with a range of forms of social exclusion and see these as directly affecting human relations with the environment. They concur that social inequality based on class affects the quality of the environment in which we live; so for example, in the affluent countries of the world, it is the poorer sections of the 'working class' that suffer unhealthy working conditions and polluted living environments. Pepper (1993: 63) argues that much socialist political action in nineteenth- and twentieth-century Western societies involves what he calls 'environmental protest' in the forms of struggles against conditions of life in factories and the ecological dislocation of mass industrialization, and the prerequisite urbanization and rural restructuring.

Pepper (1993: 3) insists that 'human rights' provided for by socialism are a prerequisite for developing a more benign relationship towards the environment. He documents the differences between 'reds' and 'greens' and argues that the way to reach an effective and practical synthesis of eco-socialism is through integrating aspects of socialism and anarchism. Most commentators on green political thought acknowledge the debt of green thinking to anarchist debates on the left (e.g. Dobson 1990). Robert Goodin (1992) sees the anarchist influence as a key difficulty in green social theory, and argues that the anarchist influence has led greens to adopt an inappropriate disorganized anti-statist politics of disparate and ineffectual social movements. The green understanding of the current environmental crisis is too important for this, however, and greens need to engage more effectively with mainstream political institutions. Pepper agrees, to some extent, but is far more accommodating of anarchism, which he sees as important in shaping deep green and social ecological analysis of human–environment relations, as well as influencing their prescriptions for change.

In writing about the possible futures of what he calls 'post-industrial socialism', André Gorz (1982) argues that the basis for socialist politics is shifting from the traditional working class to a more diverse collection of relatively marginalized and disaffected social groups. Here, his position is similar to that of Bookchin, in that it accounts for a range of social exclusions and political disaffection, and argues that new agents and actors are necessary for a radical environmental politics (see also, Scott 1990). Dobson (1990: 157) and Harvey (1990: 46) are concerned that, in adopting a more social ecological than socialist theoretical framework, eco-socialists may also adopt what they see as a more postmodern form of politics. This they see as individualist **'lifestyle' politics**, which is pluralistic, and has a post-industrial analysis that, for Pepper (1993), marginalizes the political significance of class. Marxist and trade union politics of exploitation of labour and workers' rights is not outdated, even in the richer countries of the globe, contends Pepper, and certainly is not so given the global division of labour and the exploitation of workers in underdeveloped countries by transnational corporations.

Like Bookchin, Pepper is very hostile to '**new ageism**' and spiritual aspects of ecology. I think the criticisms made of social ecology apply to eco-socialism also, however, as eco-socialists are too unreflective in their adherence to rationalist politics. Disappointingly, some eco-socialism also provides a less complex analysis than it might, in focusing on class at the expense of other forms of social difference. Social ecologists such as Roszack and Bookchin, and eco-socialists like Benton, include gender, but analyse gendered ideas and forms of environment–society relations in insufficient depth. Reading Dickens, one is left feeling that there would be many fruitful developments of the themes addressed had gender been incorporated into the analysis. Pepper manages to dismiss eco-feminism with a single quote from a newspaper, which rather hysterically misrepresents it as a 'mishmash of mysticism, morality . . . and (the) division of the world into "all that is good is female, all that is bad is male"'(Moore 1990, endorsed by Pepper 1993: 148).

Synthesizing elements of socialism, ecology and feminism, Salleh (1997) argues for an 'embodied materialism' which takes account of relations of reproduction and production and recognizes the socially constructed (false) division between humanity and 'nature' at the same time as questioning divisions and exploitations based on gender, 'race' and class. Salleh, like Vandana Shiva (1988, 1993, 1998), gives a broad account of the interconnecting webs (Plumwood 1994) of social exclusions, exploitation and oppression which shape the differing relations of human societies to their non-human environments. Not all eco-feminist accounts are able to account for such social complexity, as we will see below. Eco-feminism has been most successful where it has drawn upon the deep, social and socialist ecologies outlined above, as well as analyses of gender relations in theorizing environment–society relations. In such cases, it provides in my view the most satisfactory of 'green' theories.

Table 2.1 *Deep, social and socialist ecologies – some comparative themes*

	Ethics, nature and human nature	Approach to scientific knowledge	Conception of relationship between environment and society
Deep ecology	Humans are animals. Humans are not separate from nature, but a part of nature. Ethics must not be human-centred. All 'nature' has intrinsic value.	Some science (Western, mechanistic) is responsible for justifying and enabling environmental destruction. Systems science, such as 'Gaia' theory, understands the interdependence of the Earth's organisms and eco-systems.	Contemporary society is 'anthropocentric'. Humans as a species dominate the natural environment and are collectively responsible for damaging it. Anthropocentrism fails to appreciate that natural resources are finite.
Social ecology	Humans are a special kind of animal because they are rational and reflective beings. Ethics is inevitably human-centred because humans must attribute value to 'nature'.	Science is value free. It is shaped, however, by social forces of capitalism and the state, which means it may be complicit in environmental damage. Ecology is a revolutionary science because it sees human society and the natural world as interdependent.	The destruction of the environment is a result of intra-human domination or hierarchy. Oppression and exploitation in terms of class, caste, gender, race and by the state, leads to human exploitation of natural resources.
Eco-socialism	Humans and other animals have a specific nature or species life. Different species have different kinds of life needs. Therefore, different species need different treatment. Humans attribute ethical value and decide what such treatment might be.	Science is shaped by social forces, especially that of capitalism, which encourages the commodification of nature for profit. Science can have radical agendas, even Darwinian evolutionary science can be ecological. A range of scientific approaches is considered.	Capitalism commodifies nature and defines natural resources as objects for human use. The organization of work means that humans are alienated from nature. Social deprivation leads to environmental problems.

Eco-feminism: relations between gender and nature

Eco-feminism can be seen both as a form of 'green' theory and a form of feminism. It was first 'named' in France in 1974 by Françoise d'Eaubonne who reiterated the point made by Simone de Beauvoir in the 1950s, that women have a particular affinity with the 'natural world' due to their common exploitation by men (Mellor 1992: 51). Unlike de Beauvoir however, all the differing positions within eco-feminism have argued for the positive revaluation of the connection between women and nature that has, in Western society, been viewed negatively (Plumwood 1993: 8–9).

Victoria Davion (1994: 8) notes that there are two strands of eco-feminist theory, but I disagree that one strand is distinguished by being 'eco-feminine' and not somehow properly feminist. Mary Mellor (1992) more usefully describes the difference in terms of 'social' and 'affinity' explanations of women's relationship to nature (1992: 51–2). The division is somewhat arbitrary, as the approaches have more in common than they have differences. Both are concerned about violence against women and often also against animals, women's health (particularly the impact of reproductive technologies), the gendered division of labour in employment and the household. They have questioned forms of social inequity resulting from the exploitative relations between rich and poor countries, and see human domination of the environment as related to a worldview that justifies the domination of women. The difference between social and affinity eco-feminists is the latter's emphasis on 'neo-pagan' spirituality, and the physical bodily experiences of women as encouraging them to feel 'part of nature'.

Common to most eco-feminist positions is the use of a theory of **patriarchy** – male social domination, and the identification of patriarchy as structuring environment–society relations. Most argue that a key problem is the division between public and private spheres of life, in which the 'private' sphere is associated with nurturance, reproduction and the mundane reproduction of everyday life (cooking, cleaning and other domestic tasks) and is both devalued and associated with women. In addition, this is environmentally problematic, as men are further away than most women from the reproduction of the material conditions of life.

Plumwood argues, rather like Bookchin, that there is a connection between different forms of social domination. She sees gender, nature, race, colonialism and class as interfacing in a 'network' of oppressive 'dualisms'

(1993: 2). They exist as separate (autonomous) entities but are also mutually reinforcing in a 'web' of complex relations (1993: 194). Others have elaborated the social and historical specificity of the connection between the dominations of gender and nature by arguing that a similar worldview of Western Enlightenment rationality is at once patriarchal and anthropocentric in addition to being **Eurocentric** (Merchant 1980; Shiva 1988). This kind of eco-feminism can provide a useful bridge between social and deep ecologies. In seeing human domination of nature as a form of oppression in its own right, eco-feminists draw upon deep ecological theory, while providing a version of social ecology in which the domination of nature is also interrelated to intra-human social hierarchy and difference based on gender, race and class.

Female affinity with nature

Affinity eco-feminism is often associated with radical feminism and Western neo-paganism. In part, this connection stems from the dominance of American radical feminist thought in the emerging eco-feminist positions of the 1970s and early 1980s, and in part its negative association is due to its connection with some of the controversial figures of North American radical feminism, such as Mary Daly, Susan Griffin and Andrée Collard.

Daly's early work (1973) is an attempt to examine the male-dominated nature of Christian theology and point the way to a non-patriarchal theology, a theme which she later further develops (1979, 1984). Griffin's best-known work is *Woman and Nature* (1984) which is based around the juxtaposition of two different discourses, or ways of seeing the world. The patriarchal discourse of male-dominated 'rationalism', Christian theology, popular culture past and present, associated women with the natural world and justifies the domination of both. This is challenged by an eco-feminist discourse, which celebrates a close affinity and empathy with nature. Griffin has often been misrepresented as an 'essentialist' thinker who sets up a dualism between female culture, in which women's embodiment in menstrual and reproductive cycles, and their social role in the nurturing of children, means they are closer to nature (Hekman 1990; Ferguson 1993; Davion 1994). This theme is argued far more strongly in Andrée Collard's *Rape of the Wild* (1988) where Collard claims that:

> In patriarchy, nature, animals and women are objectified, hunted,
> invaded, colonised, owned, consumed and forced to yield and produce

(or not). This violation of the integrity of wild, spontaneous Being
is rape.

(1988: 1)

For Collard, women identify with nature through their bodies as mothers
and nurturers. The fact not all women can or do have children is not seen
as problematic, because potential motherhood is sufficient to link women
to the environment. Collard does not explain why motherhood might lead
to greater environmental consciousness and I think her work is
undeveloped and does fall prey to the essentialist critique. Griffin,
however, never suggests that men are not also embodied, but she does
argue that women are often 'closer to nature' due to cultural images and
social practices (see her preface to the Caldicott and Leyland collection,
Griffin 1983). Many social eco-feminists agree that the devalued status of
the body and animal physicality in Western culture is a key aspect of the
domination of women and nature. Plumwood (1993: 21), for example,
contends that the 'backgrounding and instrumentalisation of nature and
that of women run closely parallel', in terms not only of women's life
experience, such as domestic work, menstruation, reproduction and
childcare, but also of paid employment.

To argue, as some critics do (e.g. Segal 1987; Hekman 1990), that we
must exclude biology from social explanation, can be seen as nature-
phobic and as *socially* essentialist, as I argued in the previous chapter.
I think what affinity eco-feminists are advocating is not essentialism
but a form of 'standpoint epistemology'. This holds that people who are
socially disadvantaged have a certain perspective on the world, which
means they have a better or privileged understanding of a situation. In this
case, women associated with a disproportional amount of caring and
domestic labour can better understand the problems with our relationship
to 'nature' than those who are relatively privileged by having less
experience of nurturing work (see Hartsock 1983; Collins 1990; Harding
1991). I think this is the crux of affinity approaches, and those who
cannot accept standpoint epistemology will feel they are problematic.

Some 'affinity' eco-feminists emphasize the political significance of
women's Earth-orientated spirituality in challenging the social norms and
values of patriarchal religion. Judaism, Christianity and Islam are
criticized for their male-orientated language and (although not in the case
of the latter) imagery. Rianne Eisler (1990: 33) sees eco-feminists as
reaffirming a lost Earth-orientated religion of pagan goddess-worship, and
Carol Christ (1992: 277) argues that this legitimates female 'power and
authority'. For eco-feminist practitioners of witchcraft such as 'Starhawk'

(1982), paganism celebrates women's bodies and sexuality, and enables women to make the connection between their bodies and nature. Vandana Shiva (1988) has contended that aspects of Hinduism celebrate natural diversity and female spiritual strength, and are inspiring images for eco-feminists, although she has been criticized (Agarwal 1992) for failing to point out that other aspects endorse female subordination and the strict social hierarchy of caste. Critics have argued that merely adopting more positive myths and symbols for women will not change the reality of social institutions and practices (Biehl 1991). Charlene Spretnak (1990) has responded to such charges, plausibly perhaps, by positing that such mythologies are indirectly politicizing – they may inspire and thereby encourage political action to change social reality. This said, like deep ecology, affinity eco-feminism does tend to concentrate on our individual empathy for the 'environment'.

Plumwood is particularly concerned with what she calls a 'reversal strategy' (1993: 30) of social change. Rather than participate in the dominant culture and social institutions, affinity eco-feminists suggest subverting, resisting and replacing the dominant culture, and giving positive value to what has been previously despised and excluded. Plumwood suggests that this reinforces the initial dichotomy (1993: 33). The dualisms of humanity/nature and men/women are so closely intertwined that we must theorize a way of transcending them both. For Karen Warren (1987: 17), both women and men can 'stand with nature', but due to historical, cultural and social circumstances, they have different things to bring to the process, and this needs to be acknowledged by affinity eco-feminists. While there are difficulties with the political strategy affinity approaches offer, they provide a powerful analysis of the ways social exclusion, particularly the gendered division of labour, structures relations between society and the environment, and of the ways ideologies of the domination of 'nature' are interwoven with those which marginalize certain social groups.

Social eco-feminism: beyond the nature–culture dualism?

Social eco-feminism emphasizes the ideological links between gendered and natured domination and the social institutions and practices though which such marginalization, exploitation and oppression takes place.

In *The Death of Nature* (1980), Carolyn Merchant provides a critique of the scientific revolution of seventeenth-century Europe. She argues that both women and the natural environment were objectified and characterized as possessing similar subordinate inherent qualities, as a prerequisite for the commercial exploitation of natural resources and the social exclusion of women. These ideas are later elaborated (Merchant 1992) to focus more closely on the development of capitalism and its relation to the sexual division of labour. There are clear links between Merchant's ideas and those of eco-socialists: men are associated with commercial production and women with unpaid labour and reproduction, and commercial production is also 'alienated' from the natural world, which it pollutes and exploits.

Vandana Shiva (1988) links colonialism and racism to ecology and feminism. Shiva argues that the West has imposed its model of modernity on the rest of the globe through the ideology of scientific knowledge and the institutions and practices of industrial capitalism. This 'mal-development' has been ecologically destructive (particularly in terms of unsustainable Western agricultural practices). It has a negative impact on women in 'developing' countries, excluding them from their traditional roles in food production and also subjecting them to reproductive technologies that are invasive and inappropriate. Such 'maldevelopment' is both Western and gendered – the domination of colonized peoples, of humans over the environment, and of men over women are linked. 'Patriarchal' philosophy and science is 'dualistic', that is, it is based on notions of difference and inferiority, and justifies domination of men over women, Northern colonial powers over Southern colonized peoples, and humans over nature. Shiva combines social eco-feminism and its concern with capitalist development and trade, and male social domination as responsible for environmental damage, and spiritual eco-feminism, for she sees elements of Indian Hindu culture as fostering a closer relationship to nature.

Maria Mies similarly contends that the commodification of nature (with the patenting of agricultural crops, genetic manipulation of domestic animal reproduction) and of women's bodies (through reproductive technologies) are linked processes. Mies analyses the exploitation of women and of 'nature' in terms of social structures of domination: capitalism, patriarchy, colonialism, militarism and the state (Mies and Shiva 1993: 223–6). Mies (1986) suggests that the gender division of labour is at the core of the linked exploitations of women by men, Southern countries by the wealthy Northern states of the globe, and the natural environment by human society. The current distinction between production and reproduction has devalued women's 'essential' unpaid domestic labour. In giving birth and suckling infants, women are epistemologically privileged as they experience their bodies as productive in a stronger sense than men do (1986: 53).

In attempting to draw together analyses of intra-human hierarchy with the domination of nature, Plumwood (1993) argues for a 'master' category which is key to explaining various systems of social domination. For Bookchin, this concept is hierarchy, for Plumwood, it is 'reason' (1993: 4). Reason is a master narrative or 'story' of Western culture (1993: 196) which has been key to the construction and maintenance of oppressive relations around gender, nature, race and class. Western social thought is based on dualistic concepts, that is, pairs of opposites: culture/nature, male/female, mind/body, master/slave, civilized/primitive, production/reproduction, public/private, human/nature (1993: 43). This constructs difference in terms of power relations of domination and subordination, and Plumwood's solution is to replace dualist concepts with a non-hierarchical concept of 'difference' (1993: 59). We need to integrate differences in society (between sexes, genders, ages, species, etc.) and not see them in hierarchical terms. However, Plumwood's position is not so far from the postmodern theories of which she is so critical. As Mellor (1997: 115) points out, while she wants to talk about the 'mastery' of nature, she uses weak concepts with which to do so. Plumwood ends up with a focus on individual change, avoids talking of social domination, and reduces human domination of the environment to a 'story' and an 'identity' in which we humans are complicit. The strength of the work of those such as Merchant, Shiva and Mies is that they combine an analysis of changes in ideas about nature, gender, class and 'race' with material changes in society, such as agricultural and industrial practices, which provide us with a more comprehensive account.

Masculine science and the problem of rationalism

Both affinity and social eco-feminists tend to have a realist epistemology, while at the same time being critical of the gendered and human-dominant content of much scientific knowledge. Eco-feminists draw upon feminist critiques of science (Harding 1991; Rose 1994) in seeing the development of 'mechanistic' science as sexist in its association of women with nature. Merchant (1980) suggests that, until the scientific revolution in seventeenth-century Europe, the exploitation of nature had

Table 2.2 *Affinity and social eco-feminisms – similarities and differences*

	Social eco-feminism	Affinity eco-feminism
Nature and human nature	Human nature is historically specific. Social structures and institutions shape human nature and change the 'species-being' of humans and other species.	Patriarchal systems have led to the development of differently gendered selves. Men are likely to be separated from their biological natures as human animals.
Social organization	Society is shaped by the interrelations, both cooperative and conflictual, of various forms of social domination. These include: capitalism, patriarchy, racism and colonialism.	Society is patriarchal. Male domination is both material (physical, economic) and ideological (cultural).
Environment–society relationship	Humans collectively exploit and dominate nature, but the degree and form of that domination is shaped by social forces and divisions around gender, class, colonialism, etc.	Patriarchy is responsible for the exploitation of natural resources, the oppression of higher animal species, etc.
	Patriarchal capitalism exploits women's productive and reproductive labour.	Patriarchy assigns women caring labour. Women, inculcated with an ethic of care for others, are more likely to empathize with other species.
	Ideologies of gender difference and oppression cross-cut those of colonialism, racism and the domination of nature.	Patriarchal culture devalues both women and 'nature'. It interdefines women as closer to nature / more biological and the environment is feminized (characterized as female).
Key concepts	Patriarchy Capitalism Colonialism/imperialism Racism Materialism The sexual division of labour Production/reproduction	Patriarchy Violence Caring/empathy Culture Spirituality The sexual division of labour Reproduction/mothering

been held in check by an organic view of nature, which characterized the environment as alive, female and usually benevolent. The modern scientific worldview saw nature as a machine, and as an inert set of resources for human use, and thus sanctioned the destruction of the environment and undermined the social status of women, as healers, for example. Shiva (1988) concurs that women's traditional ways of knowing (e.g. as agriculturalists) are devalued and undermined by the Western rationalist scientific paradigm. Denying alternative ways of knowing, establishes a 'monoculture of the mind' (Shiva 1993) which is both an aspect of colonialism and a form of 'violence' against nature.

These critiques beg the question of whether eco-feminists see science itself as the problem, or the social context (white, Western, male-dominated) in which science is produced, as the problem. Postmodern feminist Susan Hekman (1990) dismisses all scientific knowledge as a cornerstone of European Enlightenment thinking, as a cultural artefact that is determined by the society that produces it. All we know about the natural world, for Hekman, is interpreted socially, and is a form of representation. Keller (1992) sees such strong relativism as highly problematic for eco-feminism, because it is based on the presumption that there is no real environment 'out there' to be explained.

Whether in its social or 'affinity' form, eco-feminism is incompatible with a strong social constructionist position. Mellor (1997) argues that eco-feminism cannot accord all agency to human society, for while social organizations and intra-human power relations affect both how we understand the world and how the world is constructed, 'the physical materiality of human life is real however it is described' (1997: 7). As I argued in Chapter 1, nature, and the scientific understanding of it, may be socially constructed and interpreted by humans, but nature also has its own agency (Soper 1995). It is a form of social (or sociological) reductionism that reduces nature to culture. I would also argue that deep, social and socialist ecologists must be ontologically realist because there has to be a 'real' (material, biological) environment 'out there' to be damaged, exploited and preserved.

All strands of ecological thinking, in many ways, stand against postmodern approaches to understanding the environment, and have often been homogenized and caricatured as biologically determinist by their postmodern critics. As feminists (even those sympathetic to postmodernism) like Diana Fuss (1989), and green social theorists (hostile to postmodernism) like Dickens (2000), argue, however, for sociologists to 'confront' biology or to take biological considerations on board, this does not mean that they are therefore somehow 'essentialist'. Both nature and society are both biologically constituted and socially constructed. What links social ecologists, eco-socialists, eco-feminists and deep ecologists, is that they all adopt a form of realist epistemology. There is a real world 'out there', there are real environmental problems with social causes and consequences, and currently society is inappropriately organized in relation to the natural environment.

Conclusion

This chapter has outlined the main schools of current 'green' thinking on the relationship between the 'environment' and human society. It has argued that all green perspectives are radical in that they assume significant change in current social practices is necessary to move us away from the current and worsening environmental crisis which they collectively think we are in.

Deep ecologists have been criticized for assuming that all human cultures are collectively and equally responsible for generating this crisis, because their theory of 'anthropocentrism' does not take account of the ways in which human societies are differentiated on grounds of caste, class, gender, ethnicity, etc. Social ecologists have a different conception of the causes of environmental crisis and argue that it is due to the forms of oppression and domination that characterize human societies and enable and encourage the exploitation of the natural environment. This is a position which might underplay the extent to which human society collectively exploits natural resources however, and some eco-feminist approaches achieve a more balanced position, which combines theories of the human domination of the environment with an analysis of social division and difference. Eco-socialists have emphasized the effects of capitalism on the environment, as have social eco-feminists. Both have seen recent developments in international trade and agro-chemical technology as problematic for marginalized groups in developed countries and for the poorer nations of the globe.

Some eco-feminists and deep ecologists see the way out of the current crisis as a re-valuation of the natural environment. They envisage this primarily through increasing human empathy and connectedness with nature, which we cultivate through our personal experience. Social ecologists and Marxist, socialist and some feminist ecologists see this approach as too individualistic, but, in general, green theory has been rather weak in theorizing how the necessary reshaping of society is to come about. Strategies for social change tend to have been left to green social movement organizations, pressure groups and political parties to develop, and it is to these we will turn in the next chapter.

Key points

- Radical green perspectives such as deep, social, socialist and feminist ecologies argue that our current social and economic practices need to change in important ways in order to avert an environmental 'crisis'.

- Deep ecology contends that social relations with the environment are political and are problematic. Contemporary societies are 'anthropocentric' or human centred and fail to recognize that nature has intrinsic value. Critics suggest this conception is simplistic, because it ignores the ways in which differences among groups of humans affect how they relate to the natural world.

- Social ecology draws on left-wing anarchism in arguing that human societies are hierarchically organized and that this has an impact on human treatment of the environment. Human domination, based on ethnicity, colonial relations, gender, class and age, is responsible for the exploitation of the planet's natural resources, and if intra-human domination ceases, so will the exploitation of nature.

- Eco-socialism emphasizes the structures and processes of capitalism in commodifying the natural world and contributing to environmental destruction and exploitation. Capitalism alienates us from our work, from 'nature' and from ourselves.

- Eco-feminism bridges some of the differences between deep, social and socialist ecologies. There are two strands of eco-feminism. 'Affinity' approaches stress the ways social gender roles enable women to have more empathy with 'nature'. 'Social eco-feminism' takes account of the ways human domination of nature interlinks with social dominations based on gender, 'race', class and other forms of social exclusion.

- While the different strands of green theory have provided a thorough and varied analysis of what is wrong with environment–society relations, it has been weak in considering the best strategies to secure change in such relations. Strategies for political and social change have been left largely to the devices of the environmental social movements.

Further reading

Andrew Dobson provides an accessible introduction in *Green Political Thought*
now in its third edition (London: Routledge, 1st edn 1990, 3rd edn 2000). For an
overview of the range of environmental perspectives see Joy Palmer (ed.), *Fifty
Key Thinkers on the Environment* (London: Routledge, 2001). Robin Attfield's
The Ethics of Environmental Concern (London: Routledge, 1983) also provides a
useful and wide-ranging introduction to the philosophical issues which are
discussed somewhat briefly in this chapter.

For the politics and philosophy of deep ecology explained by one of its
'founding' figures, see Arne Naess, *Ecology, Community and Lifestyle*
(Cambridge: Cambridge University Press, 1989). Rather more readable is Bill
Devall and George Sessions, *Deep Ecology: Living as if Nature Mattered* (Salt
Lake City: Peregrine Smith Books, 1985) and more recently, Sessions (ed.), *Deep
Ecology for the Twenty-First Century* (Boston, MA: Shambhala, 1995).

Bookchin's position is most fully elaborated in *The Ecology of Freedom* and the
most recent edition (Montreal: Black Rose Books, 1990) is of particular interest
due to its preface. For critical accounts linking social ecology to wider debates,
see Andrew Light (ed.), *Social Ecology after Bookchin* (New York: Guilford,
1998).

On relations between Marxism and ecology, see Chapter 3 of Dickens, *Society
and Nature* (Hemel Hempstead: Wheatsheaf, 1992). For material with varied and
critical positions, see Ted Benton (ed.), *The Greening of Marxism* (New York:
Guilford, 1996). On relations between socialism and green theory, Robin
Eckersley's final chapter in the aforementioned Benton collection is useful, while
more comprehensive accounts can be found in André Gorz, *Capitalism,
Socialism, Ecology* (London: Verso, 1994) and David Pepper, *Eco-socialism*
(London: Routledge, 1996). Martin Ryle's *Ecology and Socialism* (London:
Radius, 1988) provides a short and readable introduction, and also useful is Joe
Weston's collection *Red and Green* (London: Pluto, 1986).

For material on eco-feminism, there is the recent *Ecofeminist Philosophy* by
Karen Warren (Savage, MD: Rowland and Littlefield, 2000). Perhaps the best
academic overview is provided by Mary Mellor in *Feminism and Ecology*
(London: Verso, 1997). *Ecofeminism* (London: Zed, 1993), by Maria Mies and
Vandana Shiva, is a lively read, and one with a strong South perspective and
coverage.

Discussion questions

1 How might the deep ecological concept of 'intrinsic value' operate in practice? Do all aspects of the environment have equal value? How might we give equal consideration to a forest, a river, a tiger, an ant and a human baby?

2 As a species, can we be said to dominate nature?

3 Both locally and globally, how does human poverty affect the non-human environment?

4 '. . . most women already live in an alternative relation to nature' (Salleh 1997: 3). Do they? Might women have a particular insight and political perspective on the 'ecological crisis' as Salleh suggests?

Confronting environmental issues: social movements and political action

This chapter will:

- outline and evaluate different theoretical explanations of 'new' social movements
- consider the characteristics of environmentalism as a social movement
- examine the strategies of political change adopted by environmentalists
- contrast the differences between 'North' and 'South' environmental movements
- compare the different priorities and strategies of the British and German green parties in examining the relationship of environmentalism to party politics
- describe the aims and evaluate the tactics and effectiveness of a variety of social movement organizations

Introduction

The current interest in the environment and the saliency of green issues in the media and popular culture is in large part the responsibility of social movements advocating some form of the green social and political thought we looked at in Chapter 2. This chapter uses sociological theories in order to consider why 'environmentalism' emerged as a social movement, gaining increased influence and popularity from the late 1970s through to the present time.

The different ideological positions of and tactics adopted by various organizations, have led to the view that there is not a single coherent entity we can refer to as 'the environmental movement', but a huge variety of environmental 'social movement organizations'. Green movements have various political strategies for changing what they see as the problematic relationship between the environment and society, and certainly their priorities differ markedly across the regions of the globe. Some seek to alter cultural values and raise environmental consciousness through education, and may operate through 'mainstream' forms of politics, such as green parties seeking votes in '**liberal democratic**' elections. Others are attracted to 'lifestyle politics' – from the radical option of sustainable commune living, to the not so radical choice of green consumerism. More confrontational have been the differing forms of direct action such as picketing, blockading and sabotage used by environmental groups. Finally, some greens have advocated more conventional political action through 'green' political parties, perhaps pursuing alliances with social democratic or socialist parties, and engaging in constitutional pressure group activity. This has involved various tactics, such as lobbying parliamentary parties and their elected representatives, media campaigns and various charitable works.

Theoretical explanations of new social movements

Much of what sociologists have had to say about the environment has focused on environmentalism as a new social movement. Social movements can be defined as collective attempts to further common interests or goals. Social movements differ from political parties in that they aim to alter public opinion and consciousness, and emphasize change in popular culture and social organization in addition to changing policy and law, and extra-parliamentary political activity is a distinguishing feature of social movement activity.

The green movement has often been singled out as a good example of a 'new' social movement. Carl Boggs (1986) has argued that 'new social movements' developed from the 1960s onwards in the prosperous countries of the globe. In '**post-industrial**' societies, a new political climate has emerged where radical political groups are less concerned with class politics and influencing the nation state. New social movements, like environmentalism, are sceptical of conventional politics and political institutions. However, as John Scott (1990) has noted, in wealthy Western Europe, green parties abound which attempt to influence government policy, run the state or influence the making of law. There is a diverse mix of groups, and it is not necessarily the case that older social movement organizations (SMOs) within environmentalism are less radical. Globally, the range of interests, tactics and kinds of struggle are highly diverse (see Peet and Watts 1996), picking up on aspects of the inequities and forms of environmental domination we saw in the previous chapter.

A range of sociological explanations has developed to account for the emergence and form of social movement activity. The different intellectual and political cultures of America and continental Europe have led to academics seeing different things as particularly important, and developing different kinds of explanations. Crudely put, American political sociologists have tended to focus on the more political aspects of social movement activity. They want to know why protest movements develop in countries with 'liberal democratic' political systems, and how they organize in order to achieve their aims. In the European context, there has been a focus on the critical perspectives developed by social movements and an analysis of why these emerge at particular historical junctures. Luke Martell (1994: 109) has a helpful typology of the different kinds of theories, which I will use here. He identifies structural explanations, cultural explanations (both of which he sees as sociological), political explanations and action explanations (which are more 'political').

Structural explanations

Structural theories see social movements developing as a result of recent changes in economic or social structures. For this reason, some structural approaches have been referred to collectively as 'new social movement' (or 'NSM') theory. Structural theories see changes in the class structure of Western societies as able to explain the decline in class-based political movements in the 1960s and a rise in other concerns, such as environmentalism.

Alain Touraine has contributed to a number of theoretical strands within the social movements debate. With regard to structural change, he argues society is now post-industrialist or 'programmed' (1981: 29). New social movements result from the generation of new forms of social exclusion and grievance. Claus Offe (1985a) considers the politics of new social movements to be different to the politics of working-class protest, because NSMs articulate ideological positions that critically question modernity and are more concerned with cultural rather than economic transformation. An important contribution to such debates has come from Alberto Melucci (1989, 1996), who has seen contemporary societies as more complex than they were in the first half of the twentieth century. The demands of NSMs reflect this by being more diverse, and concerned with personal identity and civil liberties against intrusion by the market and the state, rather than with seeking material gain through employment and welfare systems. Scott Lash and John Urry (1987) make similar points in their analysis of the changes associated with a transition from organized to what they call 'disorganized' capitalism. The political result of this development is a more plural and diffuse range of political conflict and issues, moving away from a focus on traditional forms of class conflict in modern developed Western societies.

André Gorz (1982) considers traditional working-class political actors to be decreasingly significant, as class has declined as a form of political interest. Expanding social groups such as the middle classes are interested in non-class-based issues such as the environment. Howard Newby (1979) and Stephen Cotgrove (1982) have also seen support for environmental political issues as predominantly the concern of the middle classes, which form the rank and file of environmental organizations. However, environmentalists are part of the 'new middle classes' who are relatively poor compared to traditional middle classes, suffer relatively high job insecurity and are employed in the public sector. This thesis ignores the vast differences in age, ideology and tactics between different memberships of various environmental organizations. For example, the social position of members of direct action groups like the Hunt Saboteurs Association, and nature conservation groups such as English Nature in Britain is varied. Membership composition differs between groups and also sometimes within them. Environmentalism can also be seen to divide the middle class against itself, as middle-class managers in industries such as investment banking may be very much opposed to the demands of environmental movement organizations. Cotgrove admits that people's occupations do not necessarily determine their value system. People may choose occupations in the public sector because they are post-materialists and not interested in jobs that offer money and status, so people already have a value system that determines their occupational choice. As Stuart Lowe and Wolfgang Rüdig note (1986: 522), while the environmental movement may be class based in the sense of having a significantly middle-class membership, it is not class-driven in the traditional sense of one's class determining the content of one's interests, ideology and political affiliation.

Cultural explanations

Cultural explanations stress the significance of shifts in ideology. James Davies (1962), for example, explains new social movements in terms of rising expectations, decline in deference and the liberalization of social mores from the 1960s. These explanations overlap with those above, because they emphasize social and economic changes.

Ronald Inglehart's (1977) post-materialism approach, is perhaps the best-known cultural explanation, and contends that rising standards of living in the West means that people's material wants are now being satisfied. A survey across nine European countries suggested evidence that people were less concerned with material issues (such as employment, housing and education) in the 1970s than they were in the 1950s. Inglehart argued that people are now becoming less oriented towards materialistic and acquisitive lifestyles and see spirituality, aesthetics, political and social rights, and quality of life as more important. Inglehart (1990, 1995) has sought to defend and expand the data for the original hypothesis by arguing that post-materialism is a general trend and is not bucked by periodic recession in European economies. He further suggests that post-materialism is increasing as succeeding generations are socialized into post-materialist value systems.

Stuart Lowe and Wolfgang Rüdig (1986) agree that environmentalists are post-materialists, but do not see the necessity of a link between post-material values and affluence. Post-material values could be a result of changes in the job market, or increased media coverage of environmental issues, and, as the strength of environmental concern among some of the poorest communities of the globe testifies (see Sontheimer 1991; Peet and Watts 1996), material wealth is not a necessary basis. Lowe and Rüdig (1986: 537) point out that it is not just the changes in values that need analysing but the conditions of political constraint and capability and socio-economic circumstances in which such values emerge. As Byrne (1997: 56) notes, we need a more detailed explanation as to why post-material values may lead to different kinds of social movements in different parts of the world at different times, and in providing part of the answer to this question we can turn to political explanations.

Political explanations

Political explanations of social movement activity can be of two types, those that focus on the articulation of political grievance and their exclusion by political systems, and those focusing on how much opportunity social groups may have to articulate their views within the political system. So one kind of approach looks at political exclusions, and the other at the room for inclusion.

Charles Tilly (1978) argues that part of the explanation as to why social movements form is that groups feel their demands are not met by parties, nor are they articulated in elected assemblies, through the media or in forums such as business organizations. If authorities are antagonistic or politically repress the demands of a group, then social movements take action and may often adopt violent means. Neil Smelser (1962), however, takes the opposite approach, arguing that 'structural conduciveness' to change, in the form of open and flexible political institutions, is an important prerequisite of social movement activity.

John Scott (1990) focuses on exclusion, and argues that the aims of social movements may relate to securing representation of members or members' interests, if traditional political institutions fail to incorporate their demands and desires. Scott argues that in countries such as Sweden and the former West Germany, where the green movement has become particularly significant, the demands of environmental groups have been excluded because the demands of business are strongly reflected in policy-making in these political systems. Scott maintains that it is also significant which social groups are included in political decision-making, and which are excluded. He suggests that the 'new middle class' likely to channel their energies into social movement activity are poorer public sector workers, more disenfranchised than the traditional middle class. The ideas these new middle-class radicals articulate are also fundamentally questioning of the current political status quo (1990: 140–6). Lowe and Rüdig (1986) suggest that this exclusion from decision-making in Germany can be contrasted with greater political openness in the political systems of countries such as the UK. However, as Steve Yearley notes (1992: 88), in Germany it could be argued the state is more accommodating of green ideas as it has an electoral system that accommodates small parties and enables them to access state funding.

Action explanations

These explanations concentrate on how people organize themselves within social movements, and a particular kind of action explanation is the 'resource mobilization' theory (RMT) developed by Charles Tilly. Tilly (1978) argued that social grievances are not the main reason why social movements form. Instead, we should examine why certain grievances produce social movements and others do not. The answer, for Tilly, lies in the ability of different groups to mobilize the public and organize in order to secure their objectives. This is related to the resources a group has at its disposal, such as money, facilities, professional and other skills its supporters may have access to.

John McCarthy and Mayer Zald (1987) argue that an important resource for social movements is 'conscience constituents'. These are people and organizations who would not benefit differently from a social movement, but can be persuaded to offer their support. Some of these organizations are charitable foundations and private sector companies. So, within the environmental movement for example, an organization like the World Wide Fund for Nature encourages supporters to give money by sponsoring endangered animals for example, and operates under the status of a charitable foundation. Some social movement organizations in turn run themselves in a similar manner to a private sector company, such as Greenpeace, which has its own credit card.

As Touraine (1981) emphasizes, such approaches meet the need to involve the political actor in social movement explanations. Touraine's own model is different to RMT. He is interested in the knowledge available to actors within social movements, and how activists use that knowledge to attempt to reshape the political status quo and their own place in it. Eyerman and Jamison (1991) also emphasize the role of movement personnel in articulation of new ideological positions that focus on new issues of concern. Yearley (1992) develops such ideas in looking at how social actors articulate environmental problems rather than focusing on the problems themselves, which he considers 'relatively unimportant' (1992: 115).

The idea that, by creating knowledge, social movements define environmental problems, is part of a social constructionist approach to the environment discussed in Chapter 1. John Hannigan's (1995) 'action' theory of social movements suggests that environmental problems are defined by and through the claims-making activities of certain social groups. Hannigan identifies six 'conditions' involved in the successful construction of an environmental problem (see Box 3.1). These conditions involve the activities of social movement organizations and institutions such as the media and 'science'. There are three stages to the claims-making process. First, claims are assembled as scientists 'discover' a problem, or the media and political groups may label something as a problem. Second, claims must be presented in the public domain, which involves social movement organizations dramatizing the issue and communicating it to the public through the media. The third

---

Box 3.1

Necessary factors for the successful construction of a social problem

- Scientific authority for and validation of claims.
- Existence of 'popularizer' who can bridge environmentalism and science.
- Media attention, in which the problem is 'framed' as novel and important.
- Dramatization of the problem in symbolic and visual terms.
- Economic incentives for taking positive action.
- Emergence of an institutional sponsor who can ensure both legitimacy and continuity.

Source: Hannigan (1995: 55)

---

stage involves the 'contestation' of environmental claims, where environmental groups mobilize support of political parties and public opinion, and may try to influence government. At this stage, the claims of groups may be contested or countered by other claims, about the seriousness or nature of the problem, for example. Thus environmental groups may have to seek further legitimization for their claim, and the help of an institutional sponsor. All environmental problems, for Hannigan, are socially constructed and contested. So for those who develop action explanations, the point of interest is not the 'original source' of the movement/protest in terms of the emergence of a problem/issue/grievance and perhaps an ideology which might explain it. Rather, how movements go about popularizing their cause is the issue.

Donatella Della Porta and Mario Diani (1999: 14–15) suggest that social movement theorists are increasingly integrating aspects of the approaches outlined above, and that this has improved theoretical analysis. However, this does not overcome the drawback of all the approaches described – the failure to allow for non-human causes of environmental political concern. The objective roots to the politics of environmental social movements should not be ignored. The environmental problems which social movement organizations articulate may actually be the very reason why people support such groups in the first place. Martell (1994) sees 'action' approaches in particular as 'overly sociological', and argues that we should take support for environment movements at face value, and not assume the issues are constructed by such groups. I think this is too uncritical, and what would be helpful is an analysis that, first, considers the nature of the environmental problems, second, takes account of shifts in human social structures and, third, examines how environmental groups articulate problems through, for example, use of the mass media. This would mean we could avoid collapsing the issue itself into the perception and articulation of the issue.

Environmentalism as a social movement

Some international social movements gaining prominence in the 1960s and 1970s, such as feminism, can be seen to have undergone something of a decline, at least in terms of levels of political activism in Western societies, by the end of the twentieth century (Lovenduski and Randall 1993). Certainly we have seen an anti-feminist backlash in the media and popular culture (Faludi 1992). Environmentalism seems increasingly influential, however, in terms of the numbers of organizations making an impact on domestic and international politics, public attitudes, consumer behaviour and, thus, strategies of large corporations.

The environmental movement is particularly significant in terms of international organization. There are high levels of activism in both the developed and underdeveloped worlds, albeit that organizations may have different policy priorities (Chatterjee and Finger 1994), and there is fierce debate about the appropriate role of Northern-based movements in relation to those of the South. Governments and green organizations of developing countries have argued that it is problematic for the environment movements of the North to suggest that developing countries are not entitled to the levels of Western consumption to which they aspire. Calls for restraint on exploitation of natural resources can be seen as a form of imperialism, and such questions will be discussed in greater detail in the context of development and globalization in Chapter 5. However, as we will see later, there are many indigenous environmental movements protecting local flora and fauna. The nature of environmental problems is such that the movement has to be global in focus and organization, although inevitably this will result in intra-movement conflict. The 'environmental movement' has been characterized by incredible diversity in the theoretical stances and political approaches of its activists (Soper 1995: 254), and some suggest it is therefore inaccurate to speak of an 'environmental movement' (Yearley 1994: 156). Despite important differences between environmental protests of the North and South (which we will examine below), there is a 'movement' in which a diversity of organizations are committed to environmental protection and social change.

Specific characteristics of environmental social movements

Ron Eyerman and Andrew Jamison (1991) have argued that environmental movements are engaged in what they call 'cognitive praxis', that is, the production of innovative knowledge claims. Rather than being the organization of an interest, the green movement 'is more like a cognitive territory, a new conceptual space' (1991: 55). Eyerman and Jamison only consider a few examples to be truly social movements, and see the environment movement as one of those 'especially significant movements which redefine history, which carry the historical projects that have normally been attributed to social classes'. Yearley (1994: 155) argues that their notion of 'real' social movements as engaged in 'cognitive praxis' is far too narrow. However, I do think green intellectuals have been crucial in the movement's advancement of new knowledge claims. Peter Berger (1987: 66) has argued that environmental organizations encapsulate the political result of social structural change in Western societies with the rise of a new 'knowledge class'. Postmaterialist professionals working in education, communications, social services, etc. are in the 'business' of planning for society's non-material needs and therefore, in Berger's view, have an antagonistic relationship with the core values of capitalism. However, while this may explain the support group to which environmental movements appeal, this group is not exclusively attracted to environment movements, but also towards other less overtly politicized groupings such as alternative therapies and community arts.

These approaches emphasize the relationship between social movements and the interests of particular groups (especially class interests or other group interests that have superseded class). In concentrating on social and economic change in wealthy Western societies, however, they ignore the development and strength of environmental activism in much of the poorer regions of the South.

The more action-orientated 'American' approach has emphasized the internal dynamic of movements, their forms and abilities to organize. McCarthy and Zald (1987) talk about 'SMOs' – social movement organizations, analyse their organizational processes and institutional structures. They argue that it is inaccurate to talk of a green movement as such, concentrating on what they call the 'social movement industry'. There is an array of individual, separate environmental SMOs which collaborate over specific campaigns and share supporters, but there is competition for supporters and campaign money, for wealthy backers and for media coverage. They think environmental SMOs behave like firms, and there is evidence that groups do become more adept at using the media, increasingly professional and sophisticated over time. James Connelly and Graham Smith (1999: 79–81), for example, argue that in the case of the British SMOs of Greenpeace and Friends of the Earth (FoE), there is a pattern of increasing moderation and adoption of conventional tactics. FoE has gained increased respect among the general public and sectors of the political establishment. It still engages in direct action, but seems keen to dilute its radicalism for gains in terms of policy involvement. Greenpeace has effectively used the media as an ally, and lost an understanding of the complexity of many issues, often ignoring campaigns where media attention may be difficult to grasp and sustain. Pearce (1991: 20) argues that the Greenpeace image of 'swashbuckling success' has been partly manufactured by the media among whom it has keen supporters. We will turn our attention to the role of individual green SMOs shortly, but first it might be apposite to summarize the specific characteristics of environmentalism as a social movement, acknowledging that these are contested.

1. *Globalism*: the nature of environmental issues ensures that environmental SMOs are inevitably international in focus and in organization. Environmental problems are 'transboundary' in the terms of nation statehood. National or regionally based SMOs have seized this opportunity to cooperate therefore on certain issues.
2. *Green social thought and the promotion of an alternative value system*: while many social movement organizations engage in such activity, the green movement can be seen as radical in espousing an alternative view of society radically restructured and organized (as we saw in Chapter 2).
3. *Scientific knowledge*: despite the contestable nature of much scientific knowledge, science has been of special importance to environmentalism. The reliance on certain kinds of scientific knowledge has been crucial to the green movement, and has had an impact on the professional and bureaucratic nature of green SMOs, many of whom have their own scientific experts.

Despite differences of opinion on the nature of green SMOs themselves, and the wide range of differentially focused organizations, there are certain common characteristics, which the 'environmental movement' shares at a very general level. This said, issues and ideology are importantly divided between organizations in richer and poorer regions of the globe.

Differences between environmental movements South and North

Different regions have different fauna, flora and historical experiences and these give rise to the identification of different kinds of environmental problems, and sometimes different kinds of organizations and strategies develop to contest such problems. In Western Europe, environmental movements have challenged nuclear power and weapons, road building, habitat destruction and pollution. They have operated within the political system as campaigning green parties, as pressure groups influencing governments and public opinion, and as direct action groups. There are overlapping networks of activists often in both conservationist and radical organizations. In post-state-communist Eastern Europe, environmental movements emerged from the mid-1980s onwards, but remained limited and localized in scope and aim until the collapse of the Soviet system. The environmental problems around pollution are particularly severe in ex-state socialist countries, due to the intensity and rapidity of their industrialization under Soviet government, which took place largely in the absence of environmental regulation (see Doyle and McEachern 1998: 66–70).

In the United States, Canada, Australia and some parts of Scandinavia, environmental organizations have tended to be narrower in focus and often concentrated on wilderness issues. This is in part because in such countries, there are significant tracts of land relatively unaffected by human patterns of land use. This has, until recently, led to lack of concentration on social aspects of environmental issues and the domination of the movement by a conservationist politics (around preserving national parks for example). It is only since the 1990s that American environmental movements seem to have adopted some form of social ecological perspective. The publication of *Dumping on Dixie* by Robert Bullard (1990) stimulated debate in anti-racist, left-wing and environmental political circles. Bullard argued that the environmental movement was racist in ignoring the disproportionate environmental hazard experienced by poor black Americans (in communities for example, living in close proximity to toxic waste sites). From the early 1990s we have seen the emergence of environmental justice movements in the US for example, and the concept of **environmental justice** has been significant in environmental activism in developing countries.

The terminology of North and South is controversial in itself. This simplistic division of the developed Northern and underdeveloped or developing South, or of First and Third 'Worlds' is contested and inaccurate. There are wealthy elites in poor countries, and poor communities in the richest nations of the globe. However, as will be argued in more detail in Chapter 5, in parts of the Southern Hemisphere the links between environmental degradation and intense forms of poverty are very strong, and debates about the non-human environment are bound up with those around '**development**'.

Much of the environmental debate in the South hinges on how the process of **modernization** and development is managed and controlled, with Northern control of the development of poorer regions seen to be the problem rather than industrial development per se (Chatterjee and Finger 1994: 77). Environmental activists in the South challenge the assumptions made by some deep ecologists in the 1980s (Irvine and Ponton 1988) that Southern populations are the key environmental offenders, and argue that populations of poorer regions are disproportionately affected by environmental damage cause by the smaller but intensively consumerist North.

The Indian environmental sociologist Ramachandra Guha (1997) has argued that the 'environmentalism of the poor' differs from the environmentalism of those in wealthy parts of the globe in that it is more concerned with social justice. Just as campaigners for environmental justice in the United States have argued that 'environmental bads' are more heavily distributed among their communities, so Third World environmentalists argue that the poor regions of the globe have to deal with more severe environmental hazards than the relatively wealthy North (see Chapter 5). The notion of environmental justice also involves the defence of human rights and interests in many developing countries. The well known cases of Chipko against logging in India, the rubber tappers in Brazil to preserve the rainforest, or the Ogoni people in Nigeria against pollution, are all environmental struggles embedded in the politics of justice for dispossessed, displaced or disrupted human communities (see p. 97–9).

The radicalism of environmental protest in developing countries increased from the 1980s onwards due to the strengthening of critiques of Western-style development and the exploitation of Third World economies by transnational capital (Chatterjee and Finger 1994). The critiques of, and tension within, the theories and politics of development and the environment have led some to refer to movements for environmental justice in the Southern hemisphere as 'liberation ecologies' (Peet and Watts 1996b: 13). Richard Peet and Michael Watts (1996b: 15–20) draw on post-structural social theory (particularly the ideas of Michel Foucault) in arguing that environmentalism and environmental movements should be analysed in terms of varying 'discourses' rather than seen as unified ideologically and organizationally. For Foucault, discourses are sets of ideas carrying power relations, which operate through institutions such as governments or international organizations. Peet and Watts argue that different regions of the globe have different discourses, or ways of thinking about and acting upon the issues of 'development', modernization, the environment, etc. (1996b: 16). So different social movements in different societies have different forms of environmentalism.

While I find this quite a convincing way of thinking about the diversity of environmental movements, I would still argue that the common characteristics outlined in the section above do apply, but they can be seen to adopt regional discursive formations. Green movements operate across the globe although their aims, objectives and understanding of the world differ. They all concur that social relations (in some form or other) are inappropriately organized in relation to the natural world, and echo some aspect of the theories presented in Chapter 2. They often use scientific understandings about environmental problems, although the content of those understandings will be historically and culturally influenced depending on the locality.

Green theory and policy

While in agreement that environment–society relations must change, environmentalist strategies for political and social change are very different. Solutions to ecological problems have included limits on economic growth (Irvine and Ponton 1988), localized production and consumption (Goldsmith 1972: 86), non-violent defence (Tokar 1987: 121), redefining work (Ekins 1986; Gorz 1985) and commune living

(Bahro 1986). In the 1980s some deep ecologists controversially proposed that the world population should be reduced in line with the earth's 'carrying capacity'. Some adopted a neo-Malthusian position, that population has a tendency to outstrip food production (Hardin 1977; Bunyard and Morgan-Grenville 1987). Social, socialist and feminist ecologists argued that the cause of world hunger is a maldistribution of resources rather than an insufficiency (Lappé and Collins 1978; Bradford 1989). Environmental and Third World problems are more attributable to First World over-consumption than population levels in poorer nations (Greer 1985; Trainer 1985). Chris Cuomo (1994: 93) suggests that poverty and associated infant mortality levels, and gendered notions of male virility in certain parts of the globe promote population growth. In addition, attempts to limit the populations poorer countries have often proved counter-productive, as they are experienced as an alien form of social control (Hartman 1987). Population was a preoccupation of the 1970s and early 1980s in the green politics of Northern developed countries. Thankfully, much work in the 1990s framed the 'population issue' in its social context and argued that it is the social organization in Northern developed countries which is most problematic (e.g. Kemp and Wall 1990: 106–8).

Some argue that the solution to current environmental problems is decentralization of production and consumption, and a more locally orientated 'community' life. Kirkpatrick Sale (1985) developed the concept of '**bio-regionalism**', which has been popular with some deep greens. Sale suggests that territories are not best demarcated politically, as they are at present, but physically – by natural boundaries of water and soil type. He identifies different bioregions for ecological and human communities. The largest units are 'ecoregions', defined by common ecology such as soil typology and native flora and fauna. 'Georegions' are defined by specific geographical features, such as river basins or mountain ranges, and may be contained as a sub-unit within one ecoregion, or may span ecoregions. The smallest unit is the 'vitaregion', which operates at a local level and is the bioregion containing all the resources its inhabitants need. There is no need for environmentally damaging trade, which discriminates against developing societies, for bioregionalism is based on economic self-sufficiency and political autonomy. Vitaregions foster communal responsibility, political and social participation, and less superficial social contact. Key to these arguments is the notion of appropriate scale for human social, economic and political endeavours. For Mary Mellor (1997: 111) the assumption that all vitaregions must be able to sustain their human communities is deeply flawed given the inhospitable climate and resources of some parts of the globe. Sale assumes that ecological problems always stem from social organization and may be solved by localized living. There is no mechanism that enables his bioregional communities to think and act on a global scale, and the global character of many environmental problems cannot be solved by the local responses of self-sufficient communities. In the current context of self-interested nation states, ineffective global political organizations and multinational corporations often operating beyond political control, it is difficult to see how bioregionalism might be realized.

Most green thinkers see '**sustainability**' as a desirable goal, and argue for reduced consumption and a shift away from 'consumer society'. Consumerism boosts production levels, encouraging depletion of natural resources and contributes to pollution from the production process and consumer waste. Deep greens, in particular, have contended that Western societies must radically reduce the pressure their high consumers put on natural resources. In the last fifteen years, there has been a growth in published material on the 'greening' business. A well-known early work was John Elkington and Tom Burke's *The Green Capitalists*. Elkington and Burke (1987) assume that businesses can adjust their behaviour according to the requirements of environmental sustainability with the aid of environmental audit agencies. They argue that environmentalism is in entrepreneurs' long-term interests, as future depletion of natural resources or traffic congestion may interfere with their profit-making potential. Consumers must also make environmentally friendly purchasing choices that will, in turn, influence the supply of more environmentally friendly goods. Key to the 'green capitalist' argument is the assumption that the market can operate with motives other than competition for profit maximization in mind. Consumers are not to be concerned with 'quality' and 'value for money' but to avoid products adversely affecting 'Third World' countries or causing unnecessary waste by over-packaging, etc. (Elkington and Hailes 1988: 5).

While green consumerism may be an improvement on many contemporary formations, it is difficult to see it as a single solution to environmental problems. Many enterprises are not concerned with long-term conditions for profit maximization, as they will simply move on to the exploitation of another resource. Green consumers still may consume excessively and the point is to reduce the level of consumption per se (Irvine 1989). Green consumerism also fails to address issues of over-work in relation to high consumption, that is, that consumption priorities force people to work harder and longer than they might.

Packages that claim a single solution to the 'environmental crisis', such as population control, bioregionalism or green consumerism do not provide a solution that green groups can agree is workable, or desirable, or necessary. Green movement organizations do concur on a range of policies, however, and often adopt a mix of radical and reformist aims. These include pollution controls, encouraging people to consume less in richer societies, changing patterns of trade and ending Third World debt, encouraging public transport rather than car use, organic food production and the preservation of genetic diversity in native flora and fauna.

Movement tactics

A wide range of tactics is deployed by green social movements/social movement organizations; we will only consider a number here. I would classify the three main tactical choices made by green SMOs as 'opting in', 'opting out' and confrontation, respectively: engaging in parliamentary politics, green communes and acts of direct action, both violent (acts of sabotage, although this is arguable, and threats towards individuals and groups), and non-violent. The rest of this chapter considers these different tactics in terms of their application by various social movement organizations.

Reforming society: party politics and green consumerism

Some green activists feel that they can air their ideas in front of a wider audience should they opt to engage in traditional forms of political action. For some this is a tactical necessity in that the dissemination of green ideas transcends the need for electoral success of a specifically 'green' party. Influencing 'mainstream' parties can, in this view, prompt them to re-think certain policy objectives. In continental Europe, where coalition government is often a feature of party politics, getting a green party to secure such a proportion of the vote that it becomes a viable coalition member is a key objective. Green party politics has not been without its problems and critics, even for those SMOs whose connection with party politics lies only in external lobbying in order to influence policy priorities. Party politics is associated with careerism, electoral ambition

Table 3.1 *Forms of environmental political action*

Tactic	Ideological position	Form of organization
Informal influence	Reformist	Political parties, 'green' or mainstream. Some interest groups, e.g. the RSPCA in Britain.
Lobbying	Reformist	Pressure and interest groups on specific issues, e.g. WWF campaigns against trade in rare species/rare animal products.
Letters and petitions	Reformist or radical	Local, regional or global organizations, a gobal example being the Earth Charter Initiative.
Boycott	Radical	Local, regional or global, an international example being the Greenpeace boycott of Shell oil.
Legal action	Reformist	Most obviously, green political parties, but can be used by radical groups, e.g. as part of anti-roads protest.
Research and publicity	Reformist or radical	Various groups, e.g. FoE, WEN (Women's Environmental Network).
Education	Reformist	Various groups have specific initiatives, e.g. Earth Charter's own publications and teaching aids, WWF business education schemes.
Media stunts	Radical	Famously, Greenpeace opposing whaling.
Demonstration	Radical	Used by peace and anti-war protestors opposing nuclear weapons in the 1980s, or the Gulf War; also used by anti-vivisection campaigners in Europe, and by Reclaim the Streets protest parties and cycle rides.
Civil disobedience	Radical	Mass trespass in the UK has been used by hunt saboteurs on farmland where hunts are permitted.
Violence	Radical	Eco-sabotage can be seen as violence against property, e.g. Earth First! damage to bulldozers and other equipment in anti-roads protests. The Canadian group with their vessel *Sea Shepherd*, ramming and sinking whaling vessels. Some animal rights protesters have used threats of violence and occasionally sent incendiary devices through the post to animal experimenters.

Source: Developed from Byrne (1997:24) and Connelly and Smith (1999: 78)

and ideological compromise, and even green political actors entering party politics can find their radicalism tempered by the pressures of parliamentary electoral politics. Debates have been particularly fierce in the case of Die Grünen in Germany, whereas the British Green Party has remained largely free from ideological and tactical division. In examining the debates around ideology and tactics in parliamentary politics therefore, we will compare these two cases.

Green parties in Britain and Germany

Environmentalism in the UK is something of a paradox. Many of the campaigning organizations on environmental issues emerged first in Britain and have widespread support. Britain was the first European country to have a green party – but the Green Party of England and Wales has been the least successful in electoral terms. 'Green' political parties were a development of the environmental movement in Western liberal democracies from the 1970s. In Britain, what began as the People Party in 1973, became the Ecology Party in 1975, and the Green Party (in line with others in Europe) in 1985. While the German and French Green Parties routinely get between 5–8 per cent of the vote in general elections, and over 10 per cent in elections to the European parliament, the British party does badly. Despite a 'fluke' result of 15 per cent in the 1989 European parliamentary elections, it usually only obtains between 1–3 per cent of the vote in domestic and European elections. The British electoral system forms a significant part of the explanation for these kinds of results, but the question remains as to why there is little support for a green party, but significant involvement in environmental SMOs. In comparing green political parties in Europe, Dick Richardson and Chris Rootes (1995) argue that a key reason for the lack of success of the British Green Party is, ironically, its ideological coherence. Whereas other European parties, through being home to a variety of shades of 'green' perspective, have been able to broaden their appeal to the electorate, the British Green Party has maintained a limited and committedly radical agenda (see Box 3.2).

Die Grünen has been characterized by fierce battles over policy and elements of a radical and more moderately reformist agenda are cobbled together at election time. Die Grünen has had diverse goals encapsulating the whole range of environmental politics in Germany, and the attempts of the 1970s to achieve coalitions across social movements may have

Box 3.2

Summary of key policy objectives of the Green Party of England and Wales

- *Democratic reform*: of the electoral system, a Bill of Rights, Freedom of Information Act, regional electoral assemblies, enhance the powers of local government.

- *Foreign policy*: unilateral nuclear disarmament, leave NATO, remain in the European Union but work to strengthen the relative role of the European Parliament, to cancel Third World debt and increase aid for sustainable agricultural practice.

- *Energy*: Earth's resources are finite; therefore measures to increase energy conservation include resource taxation (not income tax), and an end to nuclear energy production.

- *Transport*: no new road building schemes. A new canal system and expansion of rail and bus public provision. New planning regulations to ensure that shops and offices are located in population centres and not out of town.

- *Economic policy*: restructuring of international trade, ecological limits on 'free' trade. Oppose multinational investment and want small domestic companies to produce for their home market. Encourage local trade and barter systems. Want to restructure banking towards smaller local companies or a regional system of independent small banks. Quality of life is key criterion for policy.

- *Human rights*: every person has a right to basic material security both physiological and psychological. An end to legal, economic and social discrimination among humans on grounds of gender, sex and sexuality, ethnicity, ability and age. Social difference must be tolerated.

- *Animal rights*: end to most vivisection licensing for weapons, household products, genetic manipulation and transplantation. End to factory farming, ban on circus use of animals, ban on hunting and bloodsports. Reform of live export and abattoir practice. Zoos to be permitted only for sanctuary and endangered species preservation.

- *Agriculture*: limit intensive use of artificial fertilizers and pesticides, subsidies for transition to organic farming, minimize transportation of food and animals, increase woodland and wilderness in rural areas, reduce size of farms, encourage diversification of crops, no genetically modified crops.

Source: www.greenparty.org.uk/policy/mfss

increased the German greens' willingness to compromise on policy. The representation of ideological difference is perhaps more of a feature of German green politics because the party membership was drawn from more politically eclectic groups than was the case in the UK. In addition, environmentalists have been less anarchic in Germany than in Britain, and far more concerned with electoral success (see Hulsberg 1988).

Eclecticism has meant dispute however, and divisions in the 1980s were apparent between more Marxist influenced eco-socialists and the pragmatic 'realos'. The realos were concerned with mobilizing a host of marginalized groups: women, gays, blacks, immigrant workers, homeless people, in addition to the working class. They advocated a programme of radical reformism and compromise with mainstream parties that would benefit potential supporters in term of policy outcomes, and argued the necessity of participating in formal politics due to the immediacy of the environmental crises (Kelly 1984: 17–23). Among the eco-socialists were a significant group of 'fundis' who emphasized ideological purity and felt party politics was of little relevance to securing 'real' change. They felt they should concentrate their energies on consciousness-raising through direct action and educational initiatives rather than electioneering. In the 1980s there emerged a Rainbow electoral alliance of left-leaning parties of which Die Grünen became a key member. Thus, in this sense the German greens have always operated more like a political party in the sense of their electoral campaigning. A series of federal programmes was developed to which all local and regional parties generally adhere. The broad policy commitment included: rights for marginalized social groups, an end to 'Third World' debt, minimum income, increased regulation of industrial and agricultural waste, division of large corporations into smaller local units, and radical decentralization in political decision-making (Hulsberg 1988).

This does not differ markedly from British Green Party policy, yet the process of electoral compromise may have made Die Grünen seem less radical to the voting public. The social movement literature would not accept so crude an explanation. Some have accounted for Die Grünen's electoral success in terms of the development of anti-capitalist sentiment among the middle classes (Eckersley, 1989). Others prefer a structural explanation that focuses on economic downturn and high unemployment, with young people voting Green due to economic policies regarding job restructuring (Bürklin 1988). Alternatively, there is 'post-materialism' (Inglehart 1990), which explains a rise in radical young voters who reject a politics of class scarcity and adopt a new green value system. A key problem with such explanations is that they do not sufficiently account for why this should apply in Germany, but not in Britain given similar social and economic circumstances (albeit different electoral systems). Alternatively, perhaps Die Grünen are simply less radical than the British greens and therefore more electable. Some of those who have left Die Grünen, such as Rudolph Bahro, felt the party became corrupt by engaging in electioneering, and had become little different in policy terms and parliamentary voting, from mainstream parties (Bahro 1986: 210–11).

While electioneering may have diluted Die Grünen's policies, once elected to the Bundestag, the Green MPs of the 1980s did act differently. They ignored traditional dress codes, and used symbolic protests in debates, such as wearing gas masks in a debate on nuclear energy. The three major parliamentary parties began to adopt a generally greener outlook – so we had the greening of the German party system alongside Die Grünen's internal division. The 1990s saw their fortunes improve with an average of 7 per cent at the polls. The term 'party-movement' has been used to describe Die Grünen (Frankland and Schoonmaker 1992), to capture the extent to which the parliamentary party is very much in and of the wider range of environmental SMOs. Political moderation has accompanied, or perhaps caused their success. Radical 'fundis' have left the party, and splinter groups, such as that led by Jutta Ditfurth, made no electoral impact in the 1990s. In entering electoral alliances realo policies predominated (Zirakzadeh 1997: 91), and, as Peter Pulzer (1995: 80) has noted, 'the Greens emerged more and more as pillars rather than subverters of the system'. While currently in coalition with the SPD (Social Democratic Party) and with the most prominent realo, Joschka Fischer, Foreign Minister, Die Grünen has undoubtedly been successful. It could be argued that this inclusion of wider and often more moderate demands broadens the possible support base of the party, but whether the greying of the greens was a price worth paying is debatable.

Some have argued that the electoral 'problem' for the British Green Party is that, from the very outset, it has adhered rigidly to a radical green position, and party members and activists have been relatively united in terms of policy (as outlined in Box 3.2) and tactics. The party has been dominated, since its inception in 1991, by radical ecologists, and some of the best-known figures, such as Sara Parkin and Jonathan Porritt, left in the 1990s, wanting a more 'practical' electoral emphasis. The party shows no signs of acceding to any such pressure. What is interesting about the British Green Party is that, in terms of its ideology, aims and demands, it is more radical than non-parliamentary groups such as

Friends of the Earth or direct action groups like Greenpeace. Many environment pressure groups do not ally themselves with the party because they feel its radicalism might jeopardize the relationship they already have with some government departments. In Britain, the Green Party, on account of its radical perspective and disdain of electoral compromise, is more like an SMO than a political party in a liberal democratic state system competing for seats in Parliament.

Whether this lack of focus on electioneering in the British case is important is a matter of debate. There are those who are sympathetic to the green cause and policies, but who despair of green political organization and tactics (Goodin 1992). Green politics in Britain is seen by some of its critics as 'lifestyle politics', based on the notion of personal transformation. Goodin argues that vegetarianism, Buddhism and paganism, commune living, alternative therapies and other lifestyle options are alienating for most potential voters. He argues for strong parties in positions of power, who will, through 'strong state' tactics, implement the green policy package. This is anathema to many British green activists, who see green action and theory as linked and based on grassroots democracy. The greens have been influential in making other political parties more environmentally aware, but this is a very limited achievement given their own estimate of environmental risks. This is perhaps the paradox of green politics: a radical message, a view of impending catastrophe, and a dislike of Stalinism as a political solution! I do think political exclusion in terms of Britain's electoral system is a factor in both the British greens' lack of enthusiasm for parliamentary politics and of parliamentary success. I do not think British environmentalists are more concerned with 'lifestyle politics' and those in Germany more overtly politicized, but that the more political nature of, for example, vegetarianism, in Britain may be a result of the relative closure of the formal political system.

Greening consumerism

Notwithstanding the undoubtedly radical strategic intentions of environmental social movement organizations, one of the main ways in which 'green' initiatives have been successfully undertaken in the West, has been through the reformist strategy of 'green' consumerism.

Green consumption is currently expanding, but with the huge range of 'greener' products to choose from, it is debatable whether what we are witnessing is merely eco-labelling and **corporate greenwashing** rather than changing industrial practices. Less cynical observers may think green consumerism is evidence of a shift in public attitudes towards the environment. In Britain, for example, Sainsburys supermarket has been given an award for its range of organic produce. Arguably, the increased stock and sale of products produced by small companies such as Belgian-based Ecover, who make washing powder and household cleaning products, may be a boon to both the eco-company and the consumer. Profit maximization may not be the key company maxim, and Ecover could fit Elkington and Burke's picture of a 'green' company. Ecover targets green consumers advertising 'ecological' products for 'people who care', and the company has been recognized by the United Nations for its achievement in developing the 'world's first' ecological factory. On a smaller scale is the British company Vegetarian Shoes, which designs and sells vegan footwear. In targeting a limited and specifically ethical market, it seems unlikely that the company has pursuit of profit as its operational maxim. The ethical market is clearly implied by the company's name, and an ethical pun contained in its advertising image: a cow clad in animal-skin-free boots (see Figure 3.1)

In the United States, the 'Simplicity Movement' can be seen as an example of a social movement adopting a radical green consumption agenda. This is reformist and individualist, and organizes itself almost as a therapy. In Seattle, for example, one can choose 'voluntary simplicity' and join a 'simplicity circle' of like-minded others who support each other through group discussion in their transition away from consumerism (Bell 1998: 62). It is difficult to see, though, how some organizations such

Figure 3.1 *'A treat for your feet if you don't eat meat!'*
(*Source*: Vegetarian Shoes)

as those advocating the plain living exemplified by Amish communities (without cars, televisions and work outside the homestead) will realize their aims. While such groups may have a 'radical' agenda, mass adoption of frugality is unlikely to be voluntary in the West. I think green consumerism can be seen as an example of the post-materialism thesis, but I also think that, in the richer parts of the globe, even if people are green consumers, they are not 'post-materialists', but 'materialists' of a different kind and degree.

Environmental social movement organizations and direct action

'Grassroots' or locally based groups of environmental campaigners engaged in direct action has been a worldwide phenomenon. Campaigns have involved different kinds of action – from tree planting to preventing soil erosion and replenishing fuel stocks in Kenya, from tree-hugging to preserve the Garhwal mountain forest in the Himalayas and tree-squatting to disrupt road-building in Britain, to sabotage of logging company equipment in North America and the Amazon Basin. In this section, we will look at various examples of groups with differing philosophies and strategies for securing social change.

Greenpeace

Greenpeace was formed in Vancouver in 1971, and has an international network of state-based organizations, with forty-two offices in thirty-two countries. Its key focus from the start was public non-violent direct action (NVDA) (Connelly and Smith 1999: 80). Its strategy is for protesters to 'bear witness' to environmental abuses by their presence at the scene and reportage through their own and national and international media. Most Greenpeace support is drawn from the regions of the affluent West, but the SMO itself considers this inevitable, given the North's responsibility for environmental destruction.

Greenpeace is best known for its graphic presentation of environmental issues, which has involved courageous, dangerous and sometimes shocking tactics, with activists scaling and occupying abandoned oil rigs, placing themselves and their inflatable boats in between whales and whaling vessels or underneath drums of toxic waste. Critics argue that

---

Box 3.3

Campaigning issues for Greenpeace

Greenpeace reserves the right to campaign on whatever environmental issues are seen as particularly salient in a particular place at a particular time. According to its current mission statement, Greenpeace aims to 'ensure the ability of the earth to nurture life in all its diversity' and currently campaigns

- for the protection of oceans and ancient forest
- for the phasing-out of fossil fuels and the promotion of renewable energies in order to stop climate change
- for the elimination of toxic chemicals
- against the release of genetically modified organisms into nature
- for nuclear disarmament and an end to nuclear contamination

Source: www.greenpeace.org.uk (12 May 2001)

Greenpeace oversimplifies environmental issues, and selects only those campaigns that are media friendly (Pearce 1991). While this may be so, it has become the organization most associated with environmental direct action. It has received international acclaim for the quality of its research, drawn on by countries in negotiations over issues such as moratoriums on whaling, bans on the dumping of radioactive waste at sea, etc. However, it has lost the support of some radical activists who see it as becoming overly bureaucratic, and being 'seduced' by the establishment. While Greenpeace has disdained links to governmental and international political bodies, it has endorsed industrial products, such as low-emission cars and non-CFC fridges, and, in Britain, through the 'ethically investing' Co-operative Bank, runs its own credit card and has been involved in endorsing 'green' electricity with the company NPower.

Greenpeace is clearly not concerned with democratic participation in the green project, but overwhelmingly with media exposure. It diverts much of its funds into the maintenance of what it openly sees as its front-line troops, professional activists whose job is to carry out the 'actions' which themselves are targeted at obtaining and sustaining media exposure. Greenpeace has a command and decision-making system which is highly secretive, elitist and almost military in nature, and the activism involves macho acts of daring for which men are often best equipped. Partly due to the issues on which it focuses, there is little if any local activism – protests in the Arctic and middle of the ocean are obviously very exclusive! The combination of media stunts, advertising and related media representation, scientific and other research, boycott, direct action by occupation and blockade, has been an effective campaigning combination. 'Action' explanations of SMOs are useful in understanding why Greenpeace is so elitist. It is concerned with producing scientific and other knowledge about issues, and securing change through media exposure. I would also argue that the effective 'exposure' of environmental issues themselves encourages critical awareness among some political institutions and publics.

Friends of the Earth (FoE)

FoE began in San Francisco in 1969, founded by David Brower, who departed from the Sierra Club, the America-based conservation group founded in 1892 that was especially interested in the preservation of biodiversity through America's national parks. Brower argued that NVDA and civil disobedience were tactically necessary for securing social change (Connelly and Smith 1999: 79), although FoE has utilized a wide variety of strategies, including moderate lobbying, company boycotts, research and media reportage.

FoE has a broad environmentalist perspective, which conceptualizes environmental problems as a consequence of the pressures placed on natural finite resources by advanced industrial capitalist society. It has a more clearly political analysis of the underlying causes of environment degradation than Greenpeace, and has held multinational corporations responsible in particular. Centralized political systems have also been seen as problematic, and FoE has stressed the importance of local democracy and the decentralization of political institutions in order to promote environmental consciousness. It encourages members to participate in local campaigns and local groups have close links with 'third force organizations'. Michael Jacobs (1996: 96–7) considers third force organizations (TFOs) vital to the analysis of social movement activity. TFOs include community organizations that are involved in initiatives such as wildlife conservation, recycling consumer 'waste', selling organic produce, and local employment and trading systems (LETS, where members barter skills, time, resources and goods, and trade through a credit note system).

Box 3.4

Policy priorities for Friends of the Earth

- protecting wildlife habitats and countryside
- encouraging sustainable agriculture
- stopping the destruction of tropical rainforests
- preventing air pollution and acid rain
- promoting waste reduction and reducing over-consumption
- forcing the clean-up of rivers and drinking water
- stopping hazardous waste dumping
- controlling dangerous chemicals, including pesticides
- reducing traffic levels and improving public transport
- stopping climate change
- protecting the ozone layer
- phasing out nuclear power, and promoting energy efficiency and renewable energy

Source: www.foe.co.uk/about_us/our_campaigns (15 March 2002)

Influenced by feminist debates and alternative activism in the 1970s, FoE has a decentralized anti-hierarchical structure. Activism is based around voluntary activity rather than professionally managed media campaigns as favoured by Greenpeace. There is no national conference to determine strategy, although there are international general meetings. It is left open for national and local groups to identify issues and appropriate actions.

FoE UK has a London-based staff and a system of local groups, and has institutional arrangements which coordinate them. The national headquarters has become highly professional, and during the 1980s shifted towards more traditional campaigning through the parliamentary lobby system (McCormick 1991). Pearce (1991) argues that this professionalism is partly a result of a commitment to academic knowledge. FoE has been effective in targeting government policy through the public inquiry system, and by producing research reports. As Connelly and Smith (1999: 79) contend, FoE is now involved in policy-making on certain issues and has gained credibility and public sympathy for its research into viable alternatives to current policies. In his survey of the green movement in the UK, Byrne (1997) gives a positive appraisal of FoE, arguing that as the Green Party is weak, and Greenpeace exclusive,

FoE offers the only chance for ecologically minded activists to participate in local political actions that do not involve law-breaking.

However, having a foot in the corridors of political power while also engaging in and supporting confrontational action has been a contradictory experience. The political inclusion of FoE has weakened its radicalism. A combination of political and action approaches might help explain the investment some key FoE staff now feel in maintaining links with the political establishment, and how they have focused their agendas on what is most effectively contested. In addition, their support among professional workers may lend some credence to either post-materialist or structural explanations of social movement activity.

Earth First!

A significant influence on the beliefs and approaches of direct action groups comes from Earth First! and its tactics of eco-sabotage or 'monkey-wrenching'. 'Monkey-wrenching' refers to the calculated sabotage of machinery involved in environmentally destructive projects. Sabotage halts the process of environmental degradation and forces companies to employ security firms and pay out for machinery repair. It also creates difficulties in securing insurance for environmentally destructive projects. The aim of such actions is to prevent companies and/or governments from pursuing certain projects. Sabotage includes spiking roads with nails, flattening tyres, putting sugar in fuel tanks, spiking trees so that nails hammered into the trunk damage saw blades. According to Andrew Dobson (1991: 225) monkey-wrenching is not mindless violence or vandalism, but is intended as a defensive strategy of resistance. However, as Goodin (1992: 135) points out, accidents and injuries to people have resulted from such activities, and, because such sabotage is sometimes dangerous and destructive, monkey-wrenchers have sometimes antagonized public opinion, and been involved in less than non-violent scuffles with authorities.

Earth First! (EF!) was launched in the United States in 1980, acquiring notoriety for its 'extreme' position, summed up in its slogan: 'No compromise in defense of Mother Earth' (see Wall 1999: 3). Criticism from the media and most of the environmental movement abounded over the statements made by the founders, Dave Foreman and Christopher Manes, who, in the early 1980s, espoused a strong anti-humanism and demonstrated an obsession with issues of 'over'-population. From the early 1990s, Foreman renounced some of the anti-humanism and rhetoric, and state repression of EF! led to links with environmental movements involved in anti-racist and animal liberation politics (Wall 1999: 5–7). Wall's analysis of Earth First! is of a network of diverse groups with various locally specific objectives, with a presence in Australia, Canada, Mexico, the UK, Ireland, Holland, Russia, Poland, South Africa, India and Egypt (1999: 170–84). In some areas, Earth First! has a symbolic presence rather than an organizational existence, as a rallying cry for various counter-culture and protest movements.

In the UK, such counter-cultural organizations have been a feature of the protests against road building (see Bryant 1996), a key focus of environmental protest in 1990s. With the tree-dwelling direct action of the Twyford Down protests came the establishment of a British-based Earth First! It does not have memberships and policy documents, but encourages protesters to take direct action to secure change and protect the earth. Earth Firsters have been involved, with other groups, in tunnelling under tree roots, occupying treetops and sabotaging bulldozers. Part of the anti-roads protest has been the 'Reclaim the Streets' actions of blockades, demonstrations, cycle protests and street parties organized by a network of activists committed to ending the 'rule of the car' by blocking main roads and causing severe traffic disruption. These actions gained much media attention and, in many cases, significant local support. After initial hostility to such direct action, and in response to pressure from its grassroots members, who were often involved in the protests, FoE moved the anti-roads protest to the top of its domestic policy agenda (see Connelly and Smith 1999: 82, 92).

In Britain, the anti-roads protesters achieved success in ensuring that British political parties are wary of instigating any new road-building initiatives for fear of the costs incurred for policing, evicting and prosecuting protesters. Wall's (1999: 139–41) explanation of the emergence of Earth First! and the eclectic range of anti-roads protest in the UK, is a political one. He sees institutional closure against green debates, the injustices of the public inquiry system, and the less activist stance of groups such as FoE and Greenpeace as vital to the emergence of this direct action. He also argues that social and economic change has been significant, with rising student numbers in the early 1990s alongside rising youth unemployment, as the activists he spoke with were generally young and either unemployed or students. I think there is a role for resource mobilization theory here as well, in terms of time and radical activist networks (available for students) for example.

Indigenous rights, development and environmental protest

The Chipko ('hugging') movement of the Garhwal mountains of the Himalayan Range in the Indian state of Uttar Pradesh gained worldwide publicity in the 1970s through the actions of mainly female Himalayan villagers hugging trees to prevent them from being felled by timber companies. Some have seen Chipko as an eco-feminist protest (Shiva 1988: 67–77; Sontheimer 1991), although they have acknowledged that it emerged among those involved in the *Sarvodaya* movement, based on the teachings of Mohandas Gandhi. The Gandhian protesters opposed the commercial use of the forest, which they sought to preserve for ecological reasons and for the subsistence of local villagers. *Sarvodaya*, meaning a movement for the 'uplift of all', had a range of programmes for locally based education, and there was a series of Gandhian ashrams for the education of women in local hill villages. My own view is that Chipko was more likely to have been Gandhian than explicitly feminist. As Bina Agarwal (1992) has noted, the presence of a large number of women does not mean a protest is feminist in inception, nor that it becomes so over time.

More recently, Chipko, which opposed Western modernization and defended the environment, has been displaced by the Uttaranchal movement. This is concerned with the creation of a separate state of Uttaranchal which could best promote Western-style economic growth. Rangan (1996: 209) argues that, in India, despite significant environmental activism, protest movements (other than those with critical Gandhian perspectives) have tended not to criticize the paradigm of Western development but to work within it. Using an action explanation, Rangan argues that local elites in this region of Uttar Pradesh created myths of Chipko as an anti-Western environmental movement to enhance their own influence, and successfully manipulated national politicians and environmentalists internationally (1996: 217).

From the 1990s, environmental campaigns in India have tended to focus on the implications of biotechnology for traditional agricultural practices, so again, there is a link between development and environmentalism. Some such movements have not involved significant numbers of women, nor had any form of feminist aim, such as the Indian Farmers' Movement. Others have had a clearly feminist orientation, such as those with the slogan 'Diverse Women for Diversity' which protested against the World Trade Organization at its Seattle conference in 1999. Until very recently (March 2002), they had had some success in ensuring that Indian patenting legislation prevented the use of genetically modified crops, but, as we will discuss in Chapter 5, movements opposing agribusiness technology have a very difficult struggle against the might of transnational corporate power and influence.

In the early 1990s, the Ogoni people in Nigeria mounted a protest against devastation of their homeland by Shell Oil. A range of organizations pledged their support for 'environmental justice' for the Ogoni, including Greenpeace and Earth First!, and 'green companies' such as the Body Shop in the UK (Wall 1999: 184). The pollution involved lakes and rivers full of dead fish, water unfit for washing in and disruption to vital field drainage systems due to oil leaks. The price of protest outside liberal democratic states is particularly high however. While Earth First! protesters in Britain padlocked themselves to the railings of the Nigerian High Commission and were arrested on minor charges, the Nigerian military government conducted a two-year campaign of displacement and torture of the Ogoni, killing around 2,000 people. Some of the leading figures in the protest were executed in 1995 on a trumped up murder charge (see Bell 1998: 20).

In addition, environmental activists may find themselves in conflict with international companies or local business organizations. In the case of the rubber-tappers union in Brazil in the 1970s and 1980s, the conflict involved the protection of the rubber-tappers' livelihoods as the Amazonian rainforest was being cleared by local ranchers and corporate land speculation. The rubber-tappers argued that they had the local knowledge of the rainforest eco-system and could make a living from the forest without destroying it. In the mid- to late 1980s, the rubber-tappers, under the charismatic leadership of 'Chico' Mendes, negotiated with the Brazilian government and the World Bank, and gained international acclaim and support, ensuring by the mid-1990s that there were significant protective measures for at least their small area of Amazonia. Yet, like the Ogoni, this was not to protect activists within the movement. Mendes was assassinated by the son of a local cattle rancher in 1988 (Palmer 2001: 303–6).

Environmental organizations in the South do have different priorities and often different principles of movement organization. As the examples from India and Nigeria indicate however, even here we cannot generalize abut South protest movements. The ability to protest and the vitality of protest movements addressing environmental problems differs in the Southern Hemisphere and may be dependent on the political context, particularly the response of the state. Political exclusion here may stimulate or curtail environmental political action, and iniquitous social and economic structures may well be the cause of environmental problems.

Conclusion

Environmentalism is a social movement with some coherence of philosophy and aim if generally and broadly defined. It is often seen by sociologists as a 'new' social movement, and associated with a range of protests from the 1970s to the present day. Given the history of environmental organization, however, it may not be as 'new' as some theorists suggest, although we are certainly witnessing an unprecedented surge of environmentalist political activity in recent times.

Sociologists have produced a range of theoretical explanations for social movement activity, some focusing on the motivations of individual actors in joining those movements and engaging in protest, others focusing on the social and economic changes that give rise to such movements. Problematically, however, such theories tend to be Eurocentric in failing to account for the circumstances of the emergence of environmentalism in developing countries. Part of the difficulty may be that current approaches stress the social origins of social movements, rather than allowing for the 'environment' itself to play a role in defining what the issues are. Environmental problems tend to be more devastating in the Southern Hemisphere, both so far as human communities and the flora and fauna are concerned. While environmental issues are defined by social movement organizations, often through their influence on the media, the problems themselves may also be real. In developing countries, the proximity of many people to those problems, combined with social inequality, may be a significant spur to environmental activism.

The environmental movement is characterized by its global perspective and activism, by its use of scientific knowledge and the generation of new kinds of social and political thought. It is, however, composed of a vast array of 'social movement organizations' which differ markedly in the perspectives to which they subscribe and the tactics which they adopt. In documenting a few such organizations, this chapter hopes to have given a flavour of the range of social movement organizations, and the ways they might influence our understanding of environmental issues by their policy priorities and the effectiveness of their campaigns.

The study of environmentalism as a social movement has been sociologically fruitful. The next chapter will try to show how the historical legacy of the discipline, in explaining the development and nature of modern industrial society, can help us understand the possible causes and consequences of many of the problems environmental social movement organizations have brought to our attention.

Key points

- Social movements are multi-faceted extra-parliamentary forms of protest, which seek social and political change of current arrangements.

- In analysing the development of social movements, American sociologists tend to emphasize the political aspects of social movement activity, while European social movement theory has been more interested in the social changes which may have led to new grievances and forms of social exclusion. They are also interested in why new radical ideologies associated with key social movements emerge in certain societies at certain times.

- Explanations for environmentalism as a social movement have varied. They include cultural change and new 'post-material' values, change in the class structure with the 'disorganization' of contemporary capitalism and the exclusion of the environment from policy-making and political concern. Such explanations fail to take account of the role the environment itself plays in causing political concern, and of important differences in the formation and struggles of environmental movements in poorer regions.

- Green consumerism is a limited strategy for securing wide-scale social change. In some specific cases, 'green companies' may have an impact, albeit one which, given their relatively small scale in a globalized capitalist market place dominated by multinational corporations, is limited and local.

- Tactics adopted by social movement organizations vary. They include the operations of 'green' political parties that are radical in policy terms and have only achieved limited electoral success. There is a contradiction between the radicalism of much of green political thought and policy aims, and the operation of mainstream political parties in liberal democracies.

- In addition to mainstream political tactics such as lobbying and petitions, the many social movement organizations have used boycotts, publicity and media stunts, demonstrations, occupations, sabotage, trespass and various other forms of civil disobedience. Different organizations have different policy priorities and emphasize different strategies to get their message across.

Further reading

On theories of social movements, the most comprehensive introduction can be found in Donatella Della Porta and Mario Diani's *Social Movements: An Introduction* (Oxford: Blackwell, 1999). For a brief summary, Chapter 4 of Paul Byrne's *Social Movements in Britain* (London: Routledge, 1997) is useful.

Andrew Jamison *et al.*, *The Making of the New Environmental Consciousness* (Edinburgh: Edinburgh University Press, 1990), provides a comparative study of environmental social movements in the Netherlands, Denmark and Sweden. The theoretical approach that emerged from the findings of this study can be found in Ron Eyerman and Jamison's *Social Movements: A Cognitive Approach* (Cambridge: Polity, 1991). While Eyerman and Jamison stress social action approaches as explanations for environmental social movements, Alan Scott's *Ideology and the New Social Movements* (London: Unwin Hyman, 1990) considers the relationship between environmental social movement organizations and political parties in the general context of liberal democratic political systems. Stephen Cotgrove, in *Catastrophe or Cornucopia* (Chichester: John Wiley and Sons, 1982) stresses broader social changes and their impact on political issues and consciousness, emphasizing the expansion of the middle classes and the development of 'post-materialism' in explaining environmental activism. Steven Yearley's chapter, 'Social Movements and Environmental Change', in Michael Redclift and Ted Benton (eds), *Social Theory and the Global Environment* (London: Routledge, 1994), provides a useful overview.

Robert Goodin's *Green Political Theory* (Cambridge: Polity, 1992) provides an interesting analysis of the relation between green thought and practice. On theories of social and political change themselves, there is a wide range of sources. The case for a bio-regional society is best seen from a reading of Kirkpatrick Sale, *Dwellers in the Land* (San Francisco: Sierra Club Books, 1985). On the population debate, Chris Cuomo's chapter in Karen Warren's collection *Ecological Feminism* (London: Routledge, 1994), provides an overview of debates. John Elkington and Tom Burke's *The Green Capitalists* (London: Gollancz, 1987) is a provocative read.

For more detail on examples of social movement activity, see Derek Wall, *Earth First! and the Anti-Roads Movement* (London: Routledge, 1999), Richard Peet and Michael Watts's *Liberation Ecologies* (London: Routledge, 1996), which has examples of movements in Latin America, West and South Africa, South and South-East Asia. Also see David Richardson and Chris Rootes (eds), *The Green Challenge: The Development of Green Parties in Europe* (London: Routledge, 1995) and Chris Rootes (ed.), *Environmental Movements: Local, National and Global* (London: Frank Cass, 1999).

It would also be useful to look at some web sites such as:

Greenpeace international www.greenpeace.org

Greenpeace UK www.greenpeace.org.uk

The Earth Charter Initiative www.earthcharter.org

Women's Environmental Network www.wen.org

The Eden Foundation www.edenfoundation.org

World Wide Fund for Nature (WWF) international website www.panda.org

WWF UK www.wwf.org.uk

Green Party of England and Wales www.greenparty.org

The European Federation of Green Parties www.dru.nl/maatschappj/politiek/groenen/europe.htm

Earthaction www.oneworld.org/earthaction

Friends of the Earth UK www.foe.co.uk

Earth First! www.hrc.wmin.ac.uk/campaigns/ef/earthfirst.html

Discussion questions

1　Are environmentalists likely to be able to persuade consumers in richer regions of the planet to consume less?

2　Will participation in parliamentary politics inevitably de-radicalize green political parties?

3　How might the 'environmentalism of the poor' differ from that of the rich?

4　Are all the theoretical explanations of social movement activity useful in explaining different aspects and forms of environmental protest? Or are some explanations more generally convincing than others?

4 ► Environment, modernity and society: industrialism and urbanization

This chapter will:

● describe how the development of industrial society in the 'West' changed social relations

● consider the impact of such changes on the non-human environment

● examine the ways in which different spaces and places are constructed, both ideologically and physically as 'wilderness', 'countryside' and 'city'

● evaluate the notion that urbanism has become 'a way of life', particularly in the richer counties of the globe, in the light of critiques of urbanism from various sociological and environmental perspectives

● compare competing ways of seeing and using the 'countryside', drawing particularly on the British example, and indicate the possible environmental impacts of changing land use

Introduction

As we have seen from the last two chapters, an important strand of environmentalism as a political ideology and social movement has been its critique of the impact of modernization. The processes that led to the development of 'modern societies' in Europe from the sixteenth century through the nineteenth, include the development of scientific knowledge and a rational worldview, urbanization and industrialization. All these processes can be seen to have fundamentally altered relations between the natural environment and human society. In terms of its origins as a discipline, sociology was an attempt to provide rational non-religious explanations of these momentous changes in Europe – an intellectual project that sought to explain modernity (Hall and Gieben 1992). Some of the sociological views on industrialization and urbanization, however, were highly critical of the human impact of such developments, and, in some cases, there are similarities and parallels that may be drawn with environmentalist arguments.

Theoretical explanations often saw the social changes in terms of 'linked antithesis', that is, contrasting pairs of concepts in which one concept described society before modernization and another described the aftermath of such processes. While with hindsight this might seem rather simplistic, it is reflective of the rapid pace and widespread extent of the changes which, for many early European sociologists, were contemporary (Lee and Newby 1991: 47). They tended to focus on the social setting wherein new and different types of social relations were apparent: the new workspace of the factory and expanding towns and cities. Key thinkers in the nineteenth century, such as Karl Marx, Emile Durkheim and Max Weber, examined modernization in human terms. Concepts such as exploitation and alienation, the **division of labour** and **anomie** were developed to explain processes such as factory work and their impact on human labourers (Giddens 1971).

While not directly referring to 'nature', sociologists were concerned with the environment, albeit built, inanimate and human-manufactured. The process of urbanization has been an important subject for sociologists from the early nineteenth century, and their readings of the urban environment have been shaped by their conceptions of appropriate relations between 'nature' and society. For Ferdinand Tönnies and Georg Simmel, urbanization was a key characteristic of the development of modern society in the West, and involved the conquest of 'nature'. In the early twentieth century, sociologists characterized the city as having a particular 'way of life' which they described as 'urbanism'. This was distinct from the 'countryside' and rural ways of life, and both were seen as different forms of managed and socially constructed environment, differentiated from the 'wilderness' which was seen as 'nature' unmediated by human society.

David Goldblatt (1996) rightly argues that classical sociologists ignored the negative consequences of Western modernity for non-human nature. However, he throws some conceptual babies out with their theoretically underdeveloped bathwater in suggesting they did not 'possess an adequate conceptual framework with which to understand the complex interactions between societies and their environments' (1996: 6). This chapter examines some sociological ideas about industrialism, urbanism, wilderness and countryside, and considers how such theories and concepts from within the legacy of classical sociology might be useful in understanding environment–society relations.

Industrial society – transcending nature?

Industrialism has been defined as a new stage of social organization where life is dominated by industrial production. This has been viewed as so phenomenal a change that it has been referred to, in Europe, as a 'revolution'. Industrialization in Britain and Europe and the global spread of Westernized development has been perhaps 'the most fundamental transformation of human life in the history of the world' (Hobsbawm 1968: 13, see also Mathias, 1969). Sociologists in the nineteenth and early twentieth centuries tended to be preoccupied with the social effects of industrialism. Such inquiry generated key areas of study such as social stratification, and the development of concepts of class and status. Sociologists have been concerned with explaining the differences between forms and historical periods of capitalist industrial development such as corporatism, monopoly capitalism, disorganized capitalism and post-industrialism. There was debate in the 1960s and 1970s about whether the way work was organized and goods produced and consumed could best be analysed in terms of a theory of capitalist society or a theory of industrial society. Much contemporary work examines the effects of a disorganized' form of capitalist society driven by consumption, and the shift to a post-industrial form of economic and social organization – a knowledge-based economy built on service provision rather than on the manufacture of goods.

Many greens would argue that these explanations operate with an uncritical acceptance of industrialism (Porritt 1986). However, I wish briefly to describe the 'classical' sociologists on the development of capitalist industrial society, as there are theories and concepts that may enhance our understanding of industrial, post-industrial and industrializing societies and their environments.

Capitalist industrialism and human alienation

Karl Marx is the foremost theorist of capitalism as a social and economic system. Marx used the term 'mode of production' to describe relationships between human producers, whom he saw as being divided by class. The mode of production changed historically, and, with the development of markets and manufacturing, feudal relations which tied people to the land came under threat, as manufacturing needed 'free' labourers. The collapse of feudalism and the emergence of capitalist industrialism meant that possession of capital (money, property and equipment) became as important as land ownership as a source of power and wealth (see Marx 1976: Chs 26–32). What is of interest here to an understanding of environment–society relations is the idea that capitalist industrialism encouraged a move of the working population away from rural to urban areas, and separated people from the land as a source of their livelihood.

For Weber, feudal agrarian society had a 'traditional' worldview that respected custom and habit. Peasants would work to reproduce their standard of living, not in order to better it. This might indicate that peasants lived relatively 'lightly' on the land and had a simple standard of living. Weber (1938) argued that capitalism was associated with cultural values of thrift, hard work, bettering oneself and accumulating profit, which were to be found in certain Protestant societies. What was distinctive about Protestantism was that it was world-affirming – it accepted the world and felt believers should partake fully in it. By comparison, in Weber's view, Catholicism is world-rejecting: the temporal world is essentially evil, but rather than trying to change it, believers should wait for a better world in the afterlife (see Lee and Newby 1991: 190). Weber saw modernizing societies as on a path of secularization. Protestant ideas would fade away, but the rational elements of the belief system would remain, such as improving one's standard of living by making money. Contemporary environmentalists are concerned with the kind of materialism Weber identified. Many argue that, in modern consumer society and societies aspiring to high levels of mass consumption, most people are concerned with improving their standard of living, which has contributed significantly to resource exploitation. In addition, Western rationalism has been held responsible for removing the magic and mystery from nature in modernity.

Durkheim is rather less critical of modernity than Weber or Marx, although he shares a concern for the experiences of people alienated by its processes. For Durkheim (1935) forms of social relationships or 'solidarity' differentiate societies. In pre-capitalist agricultural societies, 'mechanical' solidarity prevailed wherein there is a low division of labour, and people have common experiences and a collective viewpoint. In modern capitalist societies, 'organic' solidarity prevails, with an individualistic culture, specialization in the labour market and interdependence of different groups of workers. In modern society we have more freedom to develop our talents and skills, and Durkheim saw individualism as a reflection of respect for human dignity and worth

(Giddens 1987: 72). Although generally optimistic about the social effects of industrialization, Durkheim acknowledged that the increasingly complex division of labour did not always develop smoothly and could also be accompanied by 'anomie'. This occurs where industrialization is so rapid that social 'rules' cannot keep up, and life becomes unregulated. People become increasingly dissatisfied because of the gap between their aspirations and their achievements. In modern capitalist society people are always encouraged to want more than they have, and the declining influence of social institutions such as the family and organized religion encouraged competition for material achievements. This notion of the 'treadmill' of industrial production and consumption has been of interest to environmental sociologists, as we will see below, and was also an aspect of Marx's critique of capitalism.

Marx (1976) defines capitalism as existent when the owner of capital meets the seller of labour in the free market. The urban working class, dispossessed of land and property and compelled to sell their labour, are in a relationship of inherent conflict with capitalists due to the exploitative economic arrangements whereby capitalists seek to increase profits by 'squeezing labour'. Workers are not paid the full value of the goods they produce, with the extra value accruing to the capitalist as profit. As capitalist industrialism progresses, Marx thought there would be an increased tendency to replace labour power with machines to save production costs, and to subdivide tasks, making labour monotonous and repetitive. Workers would feel they had no power over their working conditions until they came to a collective understanding of the injustice of capitalist exploitation and sought to overthrow it. Throughout the twentieth century, 'industrial socialism' (Mellor 1997: 218) has developed Marx's (and Engels') concerns with the conditions of industrial workers and ignored ecological problems with industrial production. As we saw in Chapter 2, Marx was aware of human dependency on 'nature' (Marx 1976: 328), and some of Marx's ideas are pertinent to an understanding of society–nature relations; they have been developed in attempting to understand our contemporary 'alienation' from nature in industrial societies, as we will see below.

Unfortunately, in my view, from the late 1980s much European socialist thinking and research has abandoned the analysis of industrial production. Prominent figures in the European left argued that new technologies and the globalization of capitalist production had rendered 'old' Marxist analysis of capitalist production irrelevant. In these 'New Times' (Hall and Jacques 1989) we can make the best of the triumph of market capitalism by becoming radical consumers. Commodities and their images could be shaped and revamped in order to serve radical political purposes. Such arguments marginalize a critical account of environment–society relations, however, and ignore the global inequity of possible consumption, radical or not.

Industrialism and the environment

Key to the environmentalist critique of industrialism and consumer society is the notion of sustainability and sustainable development. The industrial transformation of society and its relationship with the environment is problematic. Although contested, there is evidence that the levels of pollution created by industrialism and related consumer waste (particularly in the West) is crucial in explaining a whole host of environmental problems from 'global warming' and ozone depletion to acid rain and land and water pollution. Some environmental sociologists have begun to synthesize the environmentalist critique of industrialism with Marxist and other analysis, however, and I think these analyses are significant improvements on earlier 'green' analyses of industrialism.

One of the best-known pieces of research that criticizes the assumption of economic growth and expansion of industrial production, was the *Limits to Growth* report (Meadows *et al.* 1972). This was published as a result of research commissioned by a group of ecologically concerned civil servants and industrialists who called themselves 'The Club of Rome'. Based at MIT (the Massachusetts Institute of Technology), they adopted computer-modelling techniques to make predictions about the environmental consequences of continued growth rates of five factors: population expansion, industrial production, resource depletion, food production and pollution. Their findings were that all these factors were interdependent and have reciprocal effects on each other. If one environmental problem were altered, then this would avert a human and environmental crisis, but only for a short while. The solution was their 'limits to growth thesis'. According to this, the natural resources of the planet are finite, and levels of growth in industrial societies are not compatible with the 'carrying capacity' of the earth. The idea of economic growth was defined by the report as the key problem. Critics argued that the timescale envisaged for environmental disaster is inaccurately immediate, the methodology value laden, and the Malthusian fatalism exaggerated (Cole *et al.* 1973). The original authors responded

(Meadows *et al.* 1992) that critics take their figures too literally. They were not prophetic, but intended as a general warning of what might happen. The statistics and arguments based upon them certainly read as apocalyptic however, and have been adopted by some environmentalists to support the view 'that the motorway of industrialism inevitably leads to the abyss' (Porritt 1986: 43).

Fritz Schumacher's (1973: 10–16) earlier argument in *Small is Beautiful* was more complex. Schumacher argues that it is an illusion of contemporary modernity that the 'problem' of production has been solved, and that we can continue to produce at ever-increasing rates. We need to recognize that capital is not something that we create, but something off which we live, in terms of natural planetary resources, and it is delimited by the tolerance margins of the natural world. For Schumacher, the problem with Marx's analysis of capitalism is that the labour theory of value only explains a proportion of the total capital use in producing goods for the market. 'Natural capital' consists of fossil fuels and other resources, which, unlike human labour, are not replaceable. Some environmental sociologists agree that contemporary forms of capitalist industrial production and Marx's theorization of the production process are both insensitive to the ecological context on which they are dependent. Benton (1996) has used the term 'natural limits' to describe the effects of resource scarcity on the production process, and this needs to be built into socialist analysis of capitalist production. He also argues that different kinds of industry have different kinds of environmental consequences. Industries such as forestry and agriculture that regulate the environment have different relations to their context as compared with those that appropriate environmental resources such as fishing and mining. These again differ, for those industries that produce consumer manufactures and services, although, however seemingly removed from 'nature', all processes of production involve expropriation of natural resources. We need to understand what kinds of production, organized in which ways are particularly environmentally harmful.

Peter Dickens (1996) is interested in the commodification of natural resources, and uses a key sociological concept in order to explain both this and our alienation from the environment. He argues that the division of labour, is:

> a key but neglected factor lying behind the inability of people in modern societies to adequately understand and relate to the natural

> world. Paradoxically, the more sophisticated have modern societies
> become in shaping nature, the less they have an overall understanding
> of the same nature.
>
> (Dickens 1996: 8)

For Dickens, many of our current environmental problems may be
attributed to industrial society, but this is not just because high levels of
consumption encourage increasing production and therewith polluting
industrial practices. Rather, it is the nature of industrial production itself
that is problematic. The division of labour is the main cause of people
misunderstanding their relationships with nature, and is related to the
transformation of nature into commodities for human use. The division
of labour alienates us from nature, because industrial production
disassembles and reassembles the natural world, and treats the elements
of the natural world as 'objects'. In order to substantiate his theory, he
uses case studies of modern factory farming and recent developments
in reproductive technology. Benton (1993) has also argued that the
alienation experienced by factory workers in Marx's account also can be
applied to the experience of mammals and birds in the industrial
production of factory farming.

Industrial pollution and a disregard for environmental problems
characterized the communist states of Eastern Europe, and Dickens
argues this is because the division of labour is not endemic to capitalist
production alone. Arran Gare (1996) has contended that
environmentalism flourished in the early Soviet Union, but that the
consolidation of Stalinism set the Soviet Union on the socially and
environmentally disastrous road of rapid forced industrialization. Gare
suggests that socialist environmentalism is possible and anti-ecologism is
not endemic to industrial production.

Alan Schnailberg and Kenneth Gould (1994) have used the concept of the
'treadmill of production' to describe what they see as the problematic
environmental consequences of industrial production and consumption.
Like Marx, they argue that capitalist production based on the profit
motive needs to generate supply (consumption) in order to sell and profit.
Thus the economic system has an in-built propensity to production,
(over-)consumption and pollution. Environmental abuse and social
inequity are endemic to capitalist industrialism, but are usually perceived
as inevitable and unfortunate unintended consequences. The dynamics of
this 'treadmill of production' mean that 'economic growth' and
'economic development' must continue for companies to stay afloat
(drawing on the accurate observation of Marx (1976) that the rate of profit

has a tendency to fall). Schnailberg's point (1980) was that the treadmill of industrial production is political, and that the level of consumer need in Western societies is socially constituted and sustained. Part of the construction of the treadmill of industrial production can be related to Weber's notion of the Protestant work ethic; the necessity for production is constructed through the concept of the moral virtue of hard work and individual reward (see Murphy 1994b).

These kinds of sociological analysis can help us to understand how needs and values (such as consumerism) are produced and sustained, and why capitalist industrialism is a wasteful and environmentally damaging form of productive organization which also causes social inequalities and human exploitation. Chapter 5 will develop some of these arguments in looking at the impact of European industrialism on the 'developing' areas of the world. We will now turn to another key issue affecting social relations with 'nature', looking at how humans have constructed urban and rural environments, beginning with a quick look at the possibility of a 'wilderness', minimally affected by the processes of modernity.

The world we have lost? Society and the 'wilderness'

There are different ways in which 'space' and place can be represented. Sociologists have been concerned with the representation of spaces, the economic, social and physical processes and practices which occur in different spaces, and the differing lived experience of people in places (Lefebvre 1991: 33–8). As Macnaghten and Urry (1998: 173) note, different kinds of spaces involve different practices and relations to nature. They suggest that there are four ways in which we can conceptualize such relationships. First, there is 'stewardship' of land and resources, to conserve and protect the environment for future generations. Second, land and 'natural' resources are exploited for human use; third, the environment can be 'scientized' – which involves the regulation of the environment by subjecting it to scientific investigation. Finally, the environment can be consumed as a landscape, as part of our leisure (1998: 173).

The spaces of city, town and country involve these 'ideal types' of relation. There is exploitation of agricultural land, destruction of natural environments to make way for new urban development. Stewardship of nature can be seen in some kinds of agricultural practices, such as Western organic farming and certain kinds of traditional practice in developing countries, and consumption takes place through rural property development, leisure pursuits and tourism. In parts of all the regions of the globe there are attempts to 'preserve' the wilderness, to maintain native flora and fauna often with some kind of national parks system. The scientific regulation of land space is evident in the demarcation of such national parks and sites of scientific interest.

The notion of a preserved environment is linked to the concept of some space as 'wilderness'. In hunting and gathering societies, there is no distinction between the wilderness and anything else, such as society or civilization, since all animals are wild and all land is uncultivated. The notion of the wilderness as a space, a place separate from human society, is a product of European modernity. Certainly, the Eurocentrism of preservation can be seen in the regulation of land as national parks. Ramachandra Guha (1997) has argued that the national parks approach to 'wilderness' preservation is American in origin and is based on a culturally specific understanding of an ideal environment. Forming parks was possible in the United States where the policies of genocide and containment of Amerindians on reservations had depopulated vast areas that could be preserved as 'wilderness'. In Africa and South Asia however, such swathes of depopulated land are uncommon, and the setting up of national parks in the latter half of the twentieth century has meant actively depopulating them, disrupting local economies, cultures and often environment–society relations which were relatively benign.

Short (1991) argues that this concern with wilderness preservation is a romantic perspective, and is not the one that predominates in the historical development of the 'West'. The dominant conception of 'wilderness' historically is a 'classical position', which sees human society as progressing away from the natural world, and considers the conquest of the wilderness a sign of human achievement. Carolyn Merchant (1980) charts the decline of a popular belief in animism (wherein animals, plants, rivers, seas and sky are endowed with 'life'), the rise of Christianity in Europe and the replacement of reverence for nature with fear of nature as untamed wilderness. In addition, peoples associated with 'wilderness' were characterized differently within the classical, romantic and primitivist worldviews. The former saw native peoples whom Europeans encountered through 'voyages of discovery', trade and colonial expansion as 'savages', however complex their social institutions, on the ground that they were not Christians. In the United States, the frontier has been influential in shaping American society and culture. The frontier was constructed as an area wherein the 'clash' between (white) civilization and 'nature' took place (Short 1991: 95). Thus concepts of wilderness formed a part of discourses of colonialism and racism in the eighteenth and nineteenth centuries where the peoples of countries named, demarcated and ruled by Western European states were seen as 'savages', inferior 'peasants', or, if urban and educated, still not quite 'civilized'.

Romantic radicals such as Jean-Jacques Rousseau however, had an alternative view of the 'noble savage', as freed from the corrupting constraints of 'civilization' and benefiting from a 'natural' unsullied life close to nature. In colonial discourse, the romantic vision could also be seen where colonized societies were seen as '**exotic**' with fascinating and/or more authentic social mores than those in Europe. In the late twentieth century, the notion of the 'noble savage' had credence in popular culture, academia and political protest. Native peoples, such as Amerindians, Indians of the Amazonian forests and aboriginal peoples were seen to have a closer relation to nature and to respect their environment and life within the 'natural limits' of their 'bioregion' (see for example, Brown 1970).

In the twenty-first century, there is very little 'wilderness' left, and even those places seen as wilderness, such as Antarctica, are subject to human social construction. They are transformed into images for mass consumption in television documentaries, often as spectacles for aesthetic consumption, or 'scientized' – treated as objects for conservation and scientific discoveries. Nevertheless, there is more to the notion of wilderness than ideology and representation. There are certain forms of habitat where human permanent habitation has been limited in size and impact. This is a contrast to the human-mediated environments of the city, town and 'countryside', where the physical modification and reconstruction of nature is intense.

Urbanism as a way of life

Different images of city spaces, town spaces, village spaces and 'rural' spaces are powerful social constructs in contemporary cultures. 'World' cities, such as New York, Tokyo, Bombay and Paris, are positively characterized as heterogeneous and cosmopolitan, but the 'inner city' is also associated with social problems of poverty and social exclusion, conflict and alienation in popular culture. In the pro-urban view, the city is transformative, more connected to world events, open to change and difference and less tied to tradition (Short 1991: 43). 'Regional' cities (the capitals of regions within a state, such as Bristol in Britain or Jaipur in India) and smaller towns may often be associated with images of negative parochialism, or, more positively, with a sense of 'community' or quaint aesthetics. City suburbs are viewed differently again, with associations of family life and financial security, at least in wealthy Western countries, places where people put down 'roots'. Critics of the suburban way of life, such as Lewis Mumford (1961), argued that the suburbs are culturally conformist and aesthetically homogeneous. Early feminist accounts of women's social role in the 1960s also characterized the suburbs as a place of imprisonment and oppression for women, confined to house and home (Friedan 1965; Oakley 1976). The inner city has been associated with 'unruly' and often 'uncivilized' peoples in racist cultural discourse. Stuart Hall *et al.* (1978: 161–2) contended that young black men in inner cities became 'folk devils' in 1970s Britain, scapegoats for people's fear of crime and the breakdown of social order. Likewise, in the USA at the close of nineteenth century, social problems of cities were seen in terms of race, ethnicity, difference and social exclusion, with the immigrant majorities of large cities such as New York seen as highly problematic.

Urbanization, society and 'nature'

Anti-urbanism was an important feature of nineteenth-century European sociology. For Ferdinand Tönnies, the development of urbanized industrialization is associated with a loss of community, and, by implication, a less 'natural' mode of human existence. The terms Tönnies used to characterize the social relations associated with modernization were *gemeinschaft* and *gesellschaft*. *Gemeinschaft* or 'community' described intimate social relations that characterized the pre-industrial world. *Gemeinschaft* relationships could be found where there was little social or geographical mobility, status was largely determined by birth, societies were culturally homogeneous and the family and the church were moral custodians. *Gesellschaft* meant the opposite, and loosely translated as association, it was used to describe contractual, rational and impersonal relationships. The increase in the scale and complexity of social life associated with industrial urbanization led to impersonality and relationships based on rational calculation. For Tönnies, both types of social relations could be found in urban and rural settings, but *gemeinschaft* was stronger in rural contexts.

Tönnies' concepts tended to be applied by urban sociologists in ways that grounded different types of social relations in different localities. In his essay 'The Metropolis and Mental Life', first published in 1903, Georg Simmel characterizes the city as having a deleterious psychological effect, and it is starkly contrasted to rural ways of life:

> with the tempo and multiplicity of economic, occupational and social life – it [the city] creates the sensory foundations of mental life, and in the degree of awareness necessitated by our organization as creatures dependent on differences, a deep contrast with the slower, more habitual, more smoothly flowing rhythm . . . of the small town and rural existence. Thereby the essentially intellectualistic character of the mental life of the metropolis becomes intelligible as over against that of the small town which rests more on feelings and emotional relationships.
>
> (Simmel 1991: 325)

People adapt to city living by developing a unique form of personality. Urban dwellers see market relations as the basis of social interaction and become more self-seeking in their personal relationships. There is such a degree of social difference in cities that people become satiated with variety and develop what Simmel called a 'blasé outlook'. The impersonality of the city means that urban dwellers are so estranged from fellow urbanites, that they develop a reserved persona in which there is minimal interaction between strangers. The problem with Simmel and Tönnies, however, is the assumption we are all affected in the same ways and to the same degree by urban living. They do not consider we might have an experience of urbanism and ruralism shaped and differentiated by ethnicity, gender, age, class and other forms of social difference. Behind the critique of urbanism is the notion that city living is problematic for our 'mental life' because it is in some way 'unnatural'. This point was made explicitly by Louis Wirth in 'Urbanism as a Way of Life':

> Nowhere has mankind been further removed from organic nature than under the conditions of life characteristic of great cities.
>
> (1938: 1–2)

Wirth argued that urban and rural spaces are not distinct social formations. Rather, the whole of Western culture is 'urban', and the city influences the content and organization of rural life. Wirth considers the environmental consequences of city living, such as pollution, poor housing and congestion. Using ecological arguments about animal behaviour in overcrowded and restricted habitats, he describes an 'urban jungle' full of potential conflict between its dwellers (see Lee and Newby

1991: 46–8). While contemporary environmentalists have often shied away from such an explicit use of socio-biology, the notion that urban living alters human interrelationships, as well as relations between humans and 'nature', has contemporary resonance in some green critiques of urbanism.

Greening the urban environment

A number of empirical studies undermined the conception of a difference between urban and rural ways of life and a 'loss of community' with the spread of urbanization. One of the most famous was by Michael Young and Peter Wilmott, based on Bethnal Green in East London. In the centre of one of the world's largest cities, Young and Wilmott found a community characterized by *gemeinschaft*-like relationships. 'Bethnal Greeners' were not alienated, for 'they know the faces in the crowd' (1962: 116). In America, Herbert Gans (1962) found *gemeinschaft* relations to be predominant among the Italian community in Boston whom he saw as 'urban villagers'.

Young and Wilmott's (1960) follow-up study found that more prosperous inner city dwellers sought to move to the suburbs, as the environment was better. In rich, urbanized societies, class and race affects how and where people live, and this is related to predominant concepts of an aesthetic and healthy environment. The suburbs developed in Europe from the mid- to late eighteenth century, and in the USA after1950, and symbolized the 'community' through a 'family home' and affluence gained through paid employment.

The idealization of suburbia is a gendered, class- and race-blind Western cultural construction. In many developing countries, the shifts of significant populations from rural to urban environments reflects the social changes accompanying industrialization, with the creation of work in the cities, and the environmental problems of drought and dessertification which encourage such migration. In Bombay and São Paulo, the outskirts of the city are not characterized by affluent lifestyles and cleaner air, but by severe poverty, no amenities and consequently a heavily polluted environment.

Some British planners in the 1950s were influenced by the anti-urban critique, and attempted to combine the best of the countryside and the city in 'garden cities'. A different tendency was to build upwards, but the

Table 4.1 *Narratives of place and space*

The wild	The rural	The urban
Unmodified/less modified	Regulated/managed	Human construction
Nature-centred/biocentric	human-centred production of food	Human-centred production and consumption of goods and services
Untamed	Tamed nature	Controlled nature
Dangerous	Safe	Interesting/deviant
Scientifically interesting	Technological testing	High technology
Requiring preservation	Plant/animal pests need to be controlled	Human pests need to be controlled/anti-social behaviour
Limited human habitation/ multivariate flora and fauna	Small human populations/ dense domestic animal populations	Densely populated/overcrowded
Infinite variety/or barren wilderness	Homogeneous	Human diversity, eclecticism and toleration of difference
Unpredictable human/non- human populations	Semi-predictable interactions between humanity, plants, animals, soil, weather, etc.	Semi-predictable interactions between humans and technology
The savage/'the noble savage'	Unsophisticated individual, poor peasant or good neighbour, clean living	Sophisticated individual, educated, wealthy or the sexually/morally deviant, poor, etc.
Examples: **Yosemite National Park, Antarctica, uncharted outer space**	**Examples:** **'The English village', wine-growing areas of France, Argentina, Bulgaria, etc., the feedlots of the American Midwest**	**Examples:** **New York, Tokyo, Bombay**

1960s experiment of high-rise living was disastrous, with badly constructed and poorly designed high-rise towers of apartments, often with no community facilities (Short 1991: 89). Contemporary re-developments of Western cities seem to have learned a little from previous mistakes, and high property prices and rents in the cities of the developed world suggest that urban living is as popular as ever. Many cities have been experiencing a process of gentrification in the later decades of the twentieth century (see Butler and Savage 1995), with an increased concern about the 'quality' of the urban environment. Some

Western municipalities have sought to improve the urban environment through park projects, city gardens and farms, and initiatives to cut pollution levels, such as through limits on car use in city centres. In addition, urban Europeans often wish to live in a house or apartment with some sort of outside space, whether a garden, yard, roof terrace or balcony.

Newby (1979: 29) sees this as both an example of the idealization of the countryside/outdoors, and of a blurring of the distinction between urban and rural ways of living. The urban and the rural spaces may compel different proximities to 'natural' and built 'environments', and in the affluent West, we are subject to ideals and anti-ideals of the 'city' and the 'countryside'. Such ideas are contested and often contradictory, however, and the processes of modernity are not bound within the spaces of towns and cities, but shape all 'places'. As we will see below for example, the 'countryside' ideal is undermined by the predominant use of space – food factories.

The contested countryside

The 'countryside' often becomes a symbolic 'middle way' between the 'wilderness' of unmodified nature, and the artificial built environs of the town and city. Western attitudes to the countryside have been shaped by a response known as the 'pastoral'. Within this 'myth', agricultural life is seen as more wholesome, spiritually rewarding and 'natural' than urban life. This view is the consequence of the social and economic processes of urbanization. For the past four centuries, the countryside of European and American societies has been set up as an ideal in contrast to the city with polluted air and 'loose morals' (see Wilson 1991: 26–47). The contemporary construction of the countryside is often as a less hurried, less stressful way of life, with a more organic and seasonal form of social organization.

In nineteenth-century England, the development of urban society and the enclosures, together with the industrialization of production, encouraged population migration to the towns and cities where there was employment. With the mass migration of poorer farmers and landless labourers, the de-populated 'landscape' of the countryside became an arena of farming for profit rather than subsistence, and of aristocratic leisure. In Britain, the latter persists as a minor use of countryside space, with activities such as walking and cycling being far more numerically significant. At the time of writing (March 2002) fox hunting is under concerted political attack, and the image of the countryside as an arena for such aristocratic 'sport' (see Figure 4.1) appears, I consider thankfully, to have had its day. From the second half of the twentieth century two forms of countryside use predominate: as Howard Newby (1979: 35–9) suggests, rural society is characterized by industrial farming and the more archaic practice of care and preservation. The former is the material 'reality' of rural life and the latter a 'veneer' which disguises it. Newby (1979: 19) sees the pastoral as a romantic ideology that focuses on rural aesthetics, whereas competing rationalist conceptions of the countryside focus on the economy. Short (1991) sees the environmentalist critique as part of the disillusionment with the pastoral myth, whereas Newby sees it as part of this romantic idealization of the countryside. I think the ecological critique is more hard-edged than this, and that critiques of agricultural practices, landscape preservation and organic farming are cognizant of socialist analysis of industrial capitalism. These romantic and rationalist discourses of nature correspond to what Macnaghten and Urry (1998) term 'stewardship' and 'exploitation'. These are both interlinked, however, as Urry (1995) argues; for example, the countryside is currently preserved for economic reasons and the tourist industry has emerged as an important economic feature of rural life.

An early sociological account of the political economy of the 'countryside' is provided by Marx, who suggests that the English agricultural revolution provided the prerequisites for the development of an industrial economy. The Enclosure Acts established capitalist relations of ownership and production, dramatically increasing farm size by enclosing common lands in aristocratic estates, taking away peasant rights over land and consolidating land ownership within the aristocracy (Marx 1976: 885–9). Agricultural change in eighteenth-century England led to the development of a rural proletariat reduced to selling their labour to live, and of the capitalist farmers who employ them (1976: 912). This conception is at odds with the rural idealism of those such as Tönnies, who have seen 'ruralism' as a 'way of life' standing in contrast to the urban environment, and with a far stronger sense of community, reciprocity and mutual interest (see Lee and Newby 1991: 50). Rather, it suggests there is little difference between the rural and the urban, that both spaces are structured according to capitalist industrial practice.

Further criticism of the rural ideal comes from Raymond Williams (1973: 104), who has argued that the strength of rural social networks is based on

Figure 4.1 *Hunting British foxes*
(*Source*: The League Against Cruel Sports)

the 'mutuality of the oppressed'. Poverty and economic dependence tie people to a particular locality. In the developed regions of the globe, however, it may also encourage them to leave such ties, and increasing migration of young workers to urban areas has caused concern in recent decades over a loss of 'community' in the countryside. Newby (1979) argues that villages in much of the developed world are no longer 'occupational communities' focused on farming or industrial manufactures, and that there has been an increased sense of isolation for those without social networks beyond the village.

If the romantic conception of the social constitution of 'rural life' is questionable, the presumption that rural life is somehow 'closer to nature' might also be problematic. In Britain, traditional (i.e. pre-1945) agricultural practices may have been more accommodating to indigenous wildlife, might have involved less cruelty to food animals and less chemical poisoning of land, air and water. However, agricultural land was still essentially a food factory, and the economic system was erected on the basis of social exclusion and elite privilege. The creation of the rural industrial landscape was not always a smooth and uncontested process however. Agricultural labourers resisted the pressures to urbanize and industrialize, and rights have been contested and won, at least in part, for alternative rural communities such as crofters. Elsewhere in the Northern Hemisphere, the human domination of nature and the industrialization of agriculture have not been so all encompassing. In many parts of Southern Europe, such as the Greek islands and areas of southern Italy, traditional agricultural practices have escaped much of the push towards industrialized production.

The political economy of the countryside of many European countries may be becoming more dependent on tourism. According to Urry (1995), different features of the physical environment of different countries are managed, celebrated and publicized for consumption: forests, mountains, beaches, etc. all become commodities for tourist consumption. Whether for agricultural production, or tourist consumption, the 'countryside' in Europe is subject to a particular ideological construction (see Figures 4.2, 4.3, 4.4 and 4.5).

Environmentalism has questioned chemical-based agriculture and cruelties associated with intensive livestock-rearing methods, and has provided a corrective to the popular public conceptions of the countryside as 'pastoral'. In much of Europe and America, due to the stripping out of bush and tree growth, and the use of chemical fertilizers and pesticides, agricultural land has become relatively free of 'nature'. Wild flowers, insects and small mammals have lost habitats or been killed by chemical poisoning, and larger species, which farmers see as 'pests' are often shot or hunted. Marion Shoard (1980, 1987) has argued that the clearance of agricultural land for large-scale monoculture is making the rural environment of European countries the same as that of the prairies of the north-west United States. The prairie is a 'featureless' 'food factory' (Shoard 1980: 10), and Newby noted that it is in the parts of Britain where farming is most heavily industrial capitalist that the landscape most resembles that of the prairies (1979: 214). Since the 1960s, there has been public concern and environmental lobbying on the use of chemical fertilizers and pesticides, and, most recently, over the trials of genetically modified crops, demonstrating increasingly extensive public disquiet about such agricultural innovations.

In developing countries, traditional agricultural practices are being undermined by the impact of Western industrialization and urbanization. In the poorer countries of the globe, the pattern of migration is of young men leaving rural areas to settle in the expanding towns and cities and work in industrial production, or to go to intensive plantation farms and grow food for export (Lappé and Collins 1978). Of particular concern

Figure 4.2 *Images of place I: the constructed 'wilderness' of Yosemite National Park*
(*Source*: Paul Goodey)

Figure 4.3 *Images of place II: cityscape, New York*
(*Source*: Paul Goodey)

Figure 4.4 *Images of place III: the countryside as pastoral, Ashdown forest, Sussex, UK*
(*Source*: Tony Cudworth)

Figure 4.5 *Images of place IV: the ideal leisure space? Myrtos beach, Cephalonia, Greece*
(*Source*: Tony Cudworth)

recently is the impact of Western multinational corporations on the economic and social structures of developing countries, and the ways their investment is changing agricultural practices in favour of cash crop farming for export. The development of chemical agriculture has been contested in developing countries since the 1970s, and more recently, environmental activists and farmers' organizations have opposed the policies of multinational agribusiness corporations regarding genetically modified crops. These issues will be examined in some detail in Chapter 5, which looks at the global impact of the formations of Western modernity we have discussed here.

Conclusion

The historical development of sociology has been concerned with the causes of, and social changes accompanying, the transitions to modernity in Europe. Key processes of change were urbanization and industrialization. Some of the issues of contemporary environmental concern focus on social and environmental impacts of these processes: pollution of air, water and land, traffic congestion, food safety, poor housing, ill health of humans and other animals, loss of habitat and species diversity. Sociological concepts developed to explain the human impact of industrialization and urbanization are useful in understanding historical and contemporary formations of environment–society relations. These concepts include alienation (from work, from nature, from other humans), the division of labour and relations of industrial production and consumption, 'community', and the cultural representation of space and place. The chapter has tried to show how changes in human societies have involved a fundamental alteration in relations between human societies and the environments on which they depend and within which they live.

While the social and environmental changes accompanying the transitions to modernity in Europe were historically and culturally specific, they have had an impact on a significant proportion of the globe. The practices and institutions of international trade, including that in human slaves, the economic, political and social relations of colonialism and post-colonialism, have meant that the modernization of Europe was a global phenomenon. The relations between rich and poor countries, the impact of development and underdevelopment on environment–society relations, and the uneven global distribution of environmental problems, forms the subject matter of the next chapter.

Key points

- Sociology has attempted to explain the development of 'modern' society, particularly the social changes of industrialization and urbanization in eighteenth- and nineteenth-century Europe. These developments have fundamentally altered the relationships between societies and environments.

- Theories and concepts from 'classical sociology' can help us understand environment–society relations. Weber's theory of rationalization covers the 'disenchantment' of the natural world, through scientific analysis. Marx's theory of capitalism suggests why modern economic production may be inherently wasteful and exploitative. The concept of the 'division of labour' used by Durkheim and others, involves natural phenomena as objects of production.

- Green critics have argued that industrial production involves the exploitation of 'natural capital'.

- A helpful approach to industrial production might be one which draws on both green insights regarding the exploitation of natural resources, and sociological insights on the impact of industrialism on human society.

- The urban, suburban, rural and 'wilderness' are types of space represented differently in popular culture. Different kinds of relations to nature can be seen across these social and natural spaces, and these relations change through time.

- Some sociologists argued that urbanization has led to a 'less natural' way of life, and have been concerned that the development of cities has led to a 'loss of community'.

- The continuing migration of rural populations to cities in both the developed and developing world has led to concern with environmental factors such as pollution and congestion. The developments of Western suburbia and of 'garden cities' have been attempts to produce a 'better' urban environment, with a greater element of managed 'nature' (in parks and gardens).

- The countryside has been idealized as a 'pastoral' in which life is simpler and more 'natural'. In highly industrialized countries, however, the countryside operates as a factory for the production of food and other resources, although the level of industrialization varies regionally.

- The use of rural land in affluent societies demonstrates different kinds of human–nature relations. There is domination of nature in chemical and heavily industrial agriculture, stewardship of nature through organic farming and conservation in 'national parks' or 'areas of scientific interest'. There is also the consumption of an idealized nature through leisure and tourism.

Further reading

On sociological theories of industrialization/de-industrialization and urbanization see the relevant chapters of the most recent editions of key introductory sociological texts such as: Anthony Giddens, *Sociology* (Cambridge: Polity, 1989), David Lee and Howard Newby, *The Problem of Sociology* (London: Hutchinson, 1991), in particular Chapters 2, 3 and 4, and Peter Worsley, *The New Introducing Sociology* (Harmondsworth: Penguin, 1992). Also containing useful material on the institutions and processes that shaped the development of 'modern' societies is Stuart Hall and Bram Gieben (eds), *Formations of Modernity* (Cambridge: Polity/OUP, 1992).

On industrialism, Marx's own views can be found in *Capital* (Harmondsworth: Penguin, 1976), particularly Chapters 26–32. On the de-industrialization of advanced capitalist societies see Scott Lash and John Urry, *Economies of Signs and Space* (London: Sage, 1994). Peter Dickens's use of Marxist ideas on production is again relevant here, see his *Reconstructing Nature* (London: Routledge, 1996).

On urbanization and ruralism in historical context see Raymond Williams's *The Country and the City* (London: Chatto and Windus, 1973). Also of interest, although some of the arguments are now dated, is Howard Newby's *Green and Pleasant Land? Social Change in Rural England* (Harmondsworth: Penguin, 1979). A different and more contemporary account involving material examining public views of the 'countryside' is Macnaghten and Urry, *Contested Natures* (London: Sage, 1998): see Chapters 6 and 7. John Rennie Short's *Imagined Country* (London: Routledge, 1991) has detailed analysis of representations of the rural and the urban. A comprehensive account from a Marxist perspective is David Harvey's *Consciousness and the Urban Experience* (Oxford: Basil Blackwell, 1995). John Urry's work on space and place is also interesting: see his *Consuming Places* (London: Routledge, 1995). Elizabeth Wilson gives a provocative and lively account of the different social, economic and cultural spaces and places of various 'world cities', accounting for differences of class, race and gender in *The Sphinx in the City* (London: Virago, 1991).

For material on the green critique of industrial affluent Western society, see the rather infamous Donella Meadows *et al.*'s *The Limits to Growth* (London: Earth Island, 1972) and its 1992 update and response to critics, *Beyond the Limits*. Jonathon Porritt provides a lively account in *Seeing Green* (Oxford: Blackwell, 1984); succinct yet quite complex is Fritz Schumacher's 'The Problem of Production', the opening chapter of *Small is Beautiful* (London: Abacus, 1974). More introductory material can be found in Michael Bell's *An Invitation to Environmental Sociology*, Chapters 2 and 3 (London: Sage/Pine Forge, 1988). On agriculture see Marion Shoard, *This Land is Our Land: The Struggle for Britain's Countryside* (London: Paladin, 1987).

Discussion points

1 How might industrial production alter in order to properly account for 'natural capital'?

2 How do you idealize place and space? What are your favourite/least favourite spaces?

3 How does industrialized agriculture compare to non-industrialized agriculture? In what ways, if at all, are these formations based on different conceptions of environment–society relations?

4 What is 'wilderness'?

⑤ Globalization, development and environmental change

This chapter will:

- define the concepts of development, dependency and globalization
- describe different aspects of the process of globalization
- consider the ways environmental problems are globalized
- evaluate global political responses to ecological damage
- critically discuss the role of transnational corporations as global actors, using the example of agribusiness biotechnology

Introduction

Like Chapter 4, this chapter is also concerned with the processes of modernization. The focus here, however, is mainly on the poorer countries of the globe, often located in the Southern Hemisphere, and variously referred to as the developing world, the underdeveloped world, or the Third World, usually depending on one's politics.

Western sociologists in the 1950s and 1960s tended to assume that the countries of the Southern Hemisphere were embarking on a path to development similar to that experienced by Europe and America from the late eighteenth century. A more critical school of development theory associated with the Marxist tradition emerged in the 1960s, arguing that the history of economic exploitation of the Third World needed to be acknowledged in order to account for its 'failure' to develop along European lines. This chapter begins by looking at modernization and dependency theories and how they attempted to explain the under-developed condition of the Third World. It then considers more recent incarnations of some of these debates in theories of globalization – a term that has become one of the most significant 'buzz words' of our times.

'Green' politics has adopted the mantra 'act locally, think globally' for over two decades, emphasizing the desirability of local participatory democracy and an awareness of environmental problems across the globe. This chapter is particularly concerned with global environmental problems, and their uneven regional distribution. Environmental problems have particular significance in developing countries as they can have a more devastating impact. Some sociologists have argued that there are certain economic, political and social features of such countries that make them particularly susceptible to environmental problems. The problems have included flooding, drought, land, water and air pollution, and loss of biodiversity. The causes and consequences of some of these problems illustrate the premise of this book, that environmental problems are inextricably linked to the societies in which they occur. One of the key themes of this chapter is that these 'societies' cannot be considered as discrete entities, but are linked by economic processes, elements of political integration, and the spread of cultural symbols across national and regional boundaries. They are also linked of course, by a 'global environmental commons' of air, seas and waterways, and importantly, climate.

Development, dependency and globalization

Globalization has been seen as 'an idea whose time has come', the concept social scientists should now deploy in order to understand the condition of, and changes in, society (Held *et al.* 1999: 1). The concept requires some definition, and for Malcolm Waters, globalization is:

> a social process in which the constraints of geography on social and cultural arrangements recede, and in which people are becoming increasingly aware that they are receding.
>
> (Waters 1995: 3)

The subjective element of this definition is significant. Theorists of globalization argue that the peoples of the globe are increasingly seeing themselves as part of a globalized world. Ronald Robertson (1992) argues the world is increasingly 'compressed' by our ability to travel around it quickly and to 'know' it through new forms of information and communication technology such as the Internet. Most people across the globe now have a sense of global oneness which he refers to as 'the consciousness of the world as a single place' (1992: 183). It is this subjective element of globalization theory that makes it distinct from earlier models of modernization and dependency. The notion of a world system, and of processes of modernization and dependency can help us understand how environmental problems have come to be seen as 'global' or 'globalized' in character.

Modernization, dependency and the global system

The concept of globalization has historical antecedents in the theories of modernization and dependency developed in the 1950s and 1960s. Modernization theorists generally adhered to a 'convergence hypothesis' that capitalism was to spread around the globe, accompanied by the development of liberal democracy. While this is now defunct, the perspectives that emerged as a critique are useful in thinking about globalization. Such critics, from a broadly Marxian perspective, contended that modernization was a supposedly neutral term for a highly political process: the global expansion of capitalist market relations through economic and cultural **imperialism**. Modernization theory was accused of bias towards the interests and perspectives of Western capitalist political systems.

The theories of modernization that emerged in the 1950s drew upon the approaches to social change in Europe we discussed in the previous chapter, such as the work of Max Weber and Emile Durkheim. The American sociologists Neil J. Smelser (1962) and Talcott Parsons (1960) drew on Durkheim in proposing an 'evolutionary' approach to social change where 'societies' evolve, adapting to become more functional in changed circumstances. Those influenced by Weberian sociology, particularly the notion of the 'Protestant ethic', emphasized the cultural prerequisites for 'development'. One of the best-known Weberian approaches is that of Walt Rostow (1962). Rostow proposes a theory of economic growth based on five 'stages'. His first stage is 'traditional society': agrarian societies with clear social hierarchy, little social mobility and a fatalistic attitude. The second stage is 'precondition for take-off' where the idea of possible and desirable economic growth becomes widespread. For the Third World, the stimulus to economic growth is not indigenous, but comes from the example set by the industrialized countries. The numerical expansion of entrepreneurs enables a society to 'take off', with rapid expansion of new industries, followed by a period of sustained growth – the 'drive to maturity', where communications and transport infrastructure expand. Finally we reach the 'age of mass consumption' where leading sectors of the economy shift from manufacturing industry to the production of consumer goods and services. So for Rostow, the historical experience of the Western industrialized world can be used to explain how the Third World could and should develop.

Problematically, his model did not 'fit' the social and economic changes that appeared to be happening in developing countries. Rostow was criticized for assuming that all 'traditional' societies were the same simply because they were 'not modern'. Reinhard Bendix (1967) argued that societies didn't simply move from the traditional to the modern, but that there were elements of traditional societies within modern ones. Non-Marxists such as Bendix avoided global analysis, analysing Third World development in terms of 'traditional' social structures and economic practices. Sociologists in the neo-Marxist tradition analysed the process of modernization in terms of inequality and exploitation. The Third World was not impoverished because it had failed to develop. Rather, the current social, political and economic problems which 'Third World' nations suffer are a product of their treatment by the wealthy, Northern capitalist states. Through the economic exploitation of colonialism and imperialism, Third World countries have been 'underdeveloped' to the economic benefit of Western nations.

One of the first critics of mainstream modernization approaches was André Gunder Frank (1969) who claimed it is impossible to find a 'real' society that approximates to Rostow's 'traditional' stage. Developing countries have long been incorporated into world capitalist relations, and development and **underdevelopment** are opposite sides of the same coin. He developed a model of a world 'metropolis' consisting of the governing class (primarily of the United States) with international satellites such as the elites of Third World countries. These satellites are also national metropolises with their own satellites, and so there is a chain of dependent relations from the centre of the world capitalist system to the peripheries. The key feature of this relation is that the world metropolis and national metropolises, based in the industrialized West, are accumulating resources and profits from the Third World – 'underdeveloping' it. Problematically, however, the way Frank uses the concepts of development and underdevelopment is similar to Rostow's use of tradition and modernity. Economic development does take place in the 'peripheries' and all states do not remain in the same state of underdevelopment. Critics have argued that Third World societies are too different and complex to be considered with a single model of social change (see Randall and Theobald 1985: 123) but, despite a recess in interest in global theorizing in the mid-1980s, the 1990s saw their resurgence in Marxist theories of the 'global system'.

Leslie Sklair (1994: 205) explicitly links this kind of 'global' sociology to the study of the environment. Sklair draws on the work of Immanuel Wallerstein (1974) who argued that we cannot analyse the development of isolated nation states, but need to consider the 'world system' of socio-economic development. Wallerstein divided the globe into three categories: the core, periphery and semi-periphery. The core was formed in the seventeenth century from the modernizing states of Northern Europe, which exploit (through international trade and colonial rule) the countries of the periphery, and hold back the development of the semi-periphery through relations of dependency. Sklair's work (1991, 1994) is more helpful than Wallerstein's over-general and rather vague historical account, as it uses the concept of a 'global system' to understand contemporary developments, particularly the power and influence of the transnational companies.

For Sklair, the transnational corporations (TNCs) not only exploit subordinate groups (classes in Marxist terminology); they support and benefit a 'transnational capitalist class' (1991: 38). Throughout the world there are economically dominant groups that share interests with the TNCs (1994: 206). The economic and political power of the TNCs is challenged by social movement protest, but this is of limited effectiveness as the popular culture of capitalist consumerism protects them. Critical ideas are subsumed into the dominant popular culture, for example, alternative popular music is used in advertising (1991: 42), and political movements, such as environmentalism, are being incorporated into capitalist consumerism (1994: 207, see Chapter 3). Unlike Wallerstein, Sklair sees globalization as a distinct contemporary phenomenon, but does not acknowledge that some aspects of contemporary society may be 'less' globalized than others. Sklair (1994: 208) concedes that 'global capitalism' may not be the only factor generating environmental crises, but ultimately concludes that capitalist development, and the ideology of consumption that 'drives' it, is seriously problematic for global survival. So, theories of modernization and development, dependency and world systems theory are not entirely distinct from some of the more recent approaches to globalization and have been applied to the global environment.

Defining globalization

Anthony Giddens argues that the processes of modernization in Western Europe, the impact of colonialism and neo-colonial socio-economic relations, and the emergent new social economic and political forms in globalized society are not distinct phases of 'development' but closely interconnected. Giddens adopts the view that 'modernity is inherently globalizing' (1990: 63) and defines globalization as the outcome of the transitions to modernity beginning in Europe in the seventeenth century. Critics suggest, however, that globalization is 'relatively autonomous' from modernization, and that there are a series of distinct processes that can be associated with it (Robertson 1992: 60). Whatever its origins, the processes of globalization are generally seen to be taking place in similar aspects of social life, although theorists will disagree about the extent and nature of change. Some kind of globalizing process is seen to be evident in representation, leisure, popular culture and communications, the economy, and the political system, and on this at least many sociologists agree (see Box 5.1). Such developments have changed relations between the environment and society. We are more aware of environmental issues and environmental protest has become globalized to an extent. Some have argued that economic globalization has had a particularly strong impact on both the societies and environments of the poorer nations of the globe. Before examining these specific questions, however, we need to consider how different sociologists have theorized globalization in general terms.

Box 5.1

The four processes of globalization

1. A stretching of social, political and economic activities across frontiers, such as the boundaries of nation states. This means that decisions affecting people in one part of the globe will affect people living in others.
2. An intensification of connections between politics, societies and economies.
3. A speeding up of global interactions through transport and communications technology.
4. A magnification of the impact of events. For example, because of expansion and improvements in new media, we are increasingly aware of political developments around the globe.

Source: Held *et al.* (1999: 15)

Theorizing globalization

Held *et al.* (1999) identify three 'theses' of globalization, to which theorists from varied political and theoretical perspectives subscribe in different ways. These three positions can be termed hyperglobalization, scepticism and the transformationalist thesis. I think this categorization is a very useful one, and will adopt it in describing the different perspectives here.

Hyperglobalization

'Hyperglobalizers' see globalization as a new development in the history of human social organization in which it is no longer considered appropriate for us to organize in terms of the nation states which have been the dominant political form of 'modern societies'.

The neo-liberal economist Kenechi Ohmae (1995: 5) argues that we are living in a new era in which all peoples are subject to the operation of the global market place. Globalization is de-nationalizing our economies through international networks of finance, production and trade. We now have a 'borderless economy' where the political authority of nation states is negligible. Martin Albrow (1996: 85) argues that political globalization is a new world order in which the dominant political form is liberal democracy, and that we have a common global culture in which democracy and consumer capitalism are key values. Ohmae also recognizes this shift, and suggests that the role of the politician as a national representative is now outmoded (1995: 149). The homogenization of culture is a positive development, for the economically marginalized will be drawn into a new sense of identity as global citizens, and develop an attachment to the practices of consumption and liberal democracy. For Ohmae, global institutions of governance such as the IMF (International Monetary Fund of the United Nations) and the WTO (World Trade Organization) are capable of providing the necessary regulation of the global market as we move towards the grand destination of a 'world civilization'.

Hyperglobalizers on the political left such as William Greider (1997) are concerned that the demise of the nation state and promotion of free trade will prevent any attempt to redistribute wealth through progressive taxation, or to provide adequate social services for the less well-off. In this sense, globalization can be seen to represent the 'triumph' of global capitalism, where the needs and priorities of global capital and the interests of the transnational corporations impose a neo-liberal ideological commitment and policy framework on all national governments. Whether right or left thinking, 'hyperglobalizers' see globalization as a decisive break in economic, political and social arrangements and processes, with international economic and political institutions supplanting the role and powers of nation states. We are moving inexorably towards a global, liberal 'democratic', capitalist, consumption-orientated culture. This might be overly optimistic or apocalyptic, depending one one's politics. It seems that supranational bodies may have vested interests in a 'globalizing' agenda. In the environmentalist journal *Resurgence*, for example, a WTO official is quoted, with reference to the most recent GATT (general agreement on tariffs and trade) legislation, as saying:

> it won't stop until foreigners finally start to think like Americans, act like Americans, and, most of all, shop like Americans.
>
> (*Resurgence* No. 206, 2001: 15)

Our common culture and political form may be free-market consumer capitalism, but from an environmentalist perspective, this is economically unsustainable, environmentally both impossible and disastrous, and socially iniquitous.

Scepticism

The sceptical thesis contends that globalization is no new epoch, and hardly a new series of developments. Theories of globalization are a myth, global governance is weak, and nation states still have significant power despite economic internationalization. For those on the left, globalization may be merely a less controversial term for the imposition of neoliberal economic policy through the influence of the TNCs.

In this perspective, economic interdependence is not historically unprecedented, and some argue that there may even be less economic integration at present than in the past. For example, Paul Hirst (1997) contends that, in the late nineteenth century, the Gold Standard regulated financial markets in a more rigorous way than the operations of the WTO. In addition, globalization is not the only possible development. Trends towards internationalism are counteracted and contradicted by powerful trends towards regionalism and localism (Weiss 1998). Hyperglobalizers are seen as naïve in their presumption that globalization is a result of inevitable economic tendencies. Levels of interconnections between states are intensifying, but this process is politically constructed by the policies of nation states, and, in particular, the intensification of world trade is a direct result of active United States government intervention.

Marxist and socialist theorists are particularly concerned that
'globalization' has not resulted in increased social and economic equality.
Rather, it has intensified the North/South international divide with
international trade flows benefiting the rich states of the Northern
Hemisphere and increasingly marginalizing Third World states (Hirst and
Thompson 1996). Some see the most pressing issue as the new division
of international labour. With the de-industrialization of wealthy Northern
states, TNCs are exporting jobs to the states of the Southern Hemisphere
because lack of health and safety regulations, statutory employment
rights, etc. means workers can be more heavily exploited (Krugman
1996). Alex Callinicos (Callinicos *et al.* 1994) argues that globalization is
a new phase of Western imperialism. What differentiates this 'new'
imperialism from the 'old' is that national governments are now directly
implicated in the process.

On the right, the somewhat infamous American political scientist and
one-time Pentagon advisor, Samuel Huntington, is also sceptical of the
assumption of globalization. Huntington (1996) argues that the thesis
of a common 'global culture' is a myth. He agrees with Callinicos that
inequalities between rich and poor countries are intensifying, but is only
concerned with the implications of this for international order. Economic
inequality has contributed to religious fundamentalism and aggressive
nationalism, in his view. Interestingly then, whether right-wing, liberal or
left-wing, sceptics of the globalization hypothesis collectively argue that
global governance is a project of affluent Western states which is
deliberately constructed, is not historically unprecedented, and is designed
to maintain Western economic and political supremacy.

The transformationalist position

For Manuel Castells, James Rosenau and Anthony Giddens, globalization
is a powerful force transforming contemporary society over the long term.
Unlike the hyperglobalizers and the Marxist sceptics, transformationalists
do not see an end point towards which the process of globalization is
travelling, such as the demise of the nation state, or a new form of
imperialism (Rosenau 1990). Rather, they argue that economic and
political formations are shifting. Castells (1996) suggests there is a new
international division of labour that cuts across divisions between
wealthier and poorer nations. According to him, hierarchies of wealth and
status are gradually lessening, and there is globalization of economic

production and finance. The functions of the nation state, as the building block of modern politics, are being reshaped and changed in fundamental ways. For example, the jurisdiction of an individual nation state is increasingly limited by its membership of international bodies such as the European Union, the United Nations, the WTO and GATT. Held (1991) has spoken of a 'new sovereignty regime' that is less concerned with territoriality, and where politics is increasingly focused on international networks of negotiation and policy-making. For Rosenau (1997) nation states are no longer the key forms of governing power, but, like Held, he does not see this loss of power in a zero sum way, arguing that political institutions and processes are restructuring themselves.

Giddens (1990) sees the emergence of a 'New World Order', where modern nation states are becoming reflexive in relation to their sovereignty, that is, where states now recognize that their sovereignty is not a 'fixed' phenomenon and changes over time. He argues that global bodies are assuming the function of maintaining law and order. Warfare has become globalized to the extent that it is impossible for states to wage war in blocks or alliances, as they have done in the past. Weapons of mass destruction make multi-state warfare unlikely, as does the membership of the vast majority of the states of the globe, of the United Nations. Warfare in the twenty-first century, in Giddens's view, will not become a thing of the past, but will be transformed into local and peripheral conflicts, usually of single nation states against global alliances such as NATO. Recent conflicts seem to provide some evidence to support Giddens's position, such as the Gulf War or the current American military action in Afghanistan as a response to the events of 11 September 2001.

Held *et al.* (1999) argue that we must move beyond these three positions and develop a more complex framework with which to analyse 'globalization'. This framework combines aspects of the three theses in looking at different historical forms of globalization and the different kinds of organizations and processes that can be identified within them. Four different types of globalization are defined ('thick', 'diffused', 'expansive', 'thin'), differentiated by their extensive regional coverage, intensity of their 'connectedness', the speed of their processes and their impact (1999: 25). Certainly more complex theorizing might result in a better fit between abstract ideas and changes in 'the real world'. Globalization is a contested concept, but one which may be useful in thinking about relations between societies and their environments. There are fierce debates among activists and academics over the extent to which the various processes of globalization have a negative impact on the natural world and the human communities that interact with it. Before looking at these debates and issues, however, we will consider the different areas of social life globalization is said to affect.

The process of globalization

The processes of globalization operate in different areas of social life. Economic globalization refers to increasing levels of transnational trade and means of economic regulation. Social aspects of the globalizing process include the increasingly common cultural symbols with which different societies are familiar. Political globalization concerns the global quality of political issues, such as the environment, the management of these issues through global institutions, and the ways protest groups and non-governmental organizations campaign on a global level.

Social change

For many sociologists, the cultural impact of globalization is the most interesting and/or significant arena in which change can be seen to be taking place. For Waters (1995) the key globalizing process is 'symbolic exchange'. Certain kinds of cultural symbols are easily produced, reproduced and transported globally. Cultural aspects of social globalizing societies include: expansion of travel and tourism, the development of new infrastructures of telecommunications, the spread of the Anglo-American music industry, the impact of the film industry from Hollywood to 'Bollywood', and the growth and influence of European film-making, the global images of advertising and fashion, and the global influence of CNN. It is suggested that we are moving towards a common culture in which we are all familiar with the same cultural symbols at a certain point in time. As Steven Yearley describes it, writing in 1996:

> To some extent these cultures are transplanting indigenous traditions and enthusiasm, but even when they are not, they are adding a level of global familiarity. Madonna and Maradona are a safe bet as topics of conversation more or less wherever one travels.
>
> (Yearley 1996: 6)

I am rather sceptical of Yearley's rather flimsy examples (especially the second!), and critics such as Callinicos (Callinicos *et al.* 1994) see the spread of global cultural symbols as a form of American cultural imperialism. The cultural images that have global significance are largely associated with American-based transnational corporations and their brand names such as McDonald's, Pepsi and Coca-Cola. Arjun Appadurai (1990), while acknowledging the neo-colonial causes and content of the global spread of cultural mores and images, argues that the common culture of globalized society is not exclusively that of colonizing powers or white, Western, affluent societies. In Britain, for example, he argues that the colonial legacy and establishment of a significant South Asian community has meant that British culture has become more eclectic. Eclecticism in food, drink, dress and cultural imagery may be a feature of the globalization of society, but it is unevenly spread across states in the same region, and regions within states. In their criticisms of consumerism, the role of the TNCs and the consequences of international trade, environmentalists of most hues tend to be sceptical of the extent of intercultural exchange and its democratizing possibilities.

Economic factors

International trade has increased in level and volume over the last century. Particularly since 1945, however, some scholars of globalization contend that both nation states and individual companies have become enmeshed in a global trading system that has undermined the sovereignty and authority of individual nation states. Related to this development is a shift in patterns of finance, where capitalist markets operate on a global basis and the economic fate of whole countries or regions can fall prey to the fears, real or often imaginary, of investors in the international money markets.

There is some evidence that the production of goods themselves has become more globalized. In an examination of the motor industry, Wyn Grant (1993) notes that car parts will be manufactured in different areas of the globe, and the car assembled in a different continent from where those parts are manufactured. A car is no longer the product of a particular country, companies are becoming stateless and are decreasingly likely to have a home base:

> The company no longer sees itself as being based in one country, but as operating globally. The headquarters . . . could be located anywhere. There is no longer loyalty to a particular country which is seen as the 'home country' but rather to a firm which orientates itself to the global economy.
>
> (Grant 1993: 61)

Scott Lash and John Urry (1994) link the economic and political processes of globalization. In 1987, they made the case that contemporary capitalism was 'disorganized' to the extent that it is now consumption-orientated and its workforce is far more mobile. With developments in communications technology and increased speed of travel, they argue that what is most significant now is how goods and their consumers are represented, and representations of objects are the most mobile of commodities. Economic globalization, with the increased mobility of goods and people across the 'space' of the globe, means that political practices inevitably become transnational, and we are seeing the development of globalized bureaucracies and an overall decline in nation state power and authority.

Political institutions and issues

In the aftermath of the Second World War some have argued, we have also seen the globalization of politics, with the setting up of global political institutions and the decline of political territoriality, with the nation state having its authority increasingly compromised. Gareth Porter and Janet Brown (1991) speak of institutions such as the United Nations as institutions of 'global governance' and argue that, in certain areas of policy-making, such as the environment, global institutions should have overriding policy-making authority. Whether the UN might ever possess such authority is debatable. More certain is the existence of increasingly important webs of international agreements and legislation which bind the operations of individual nation states such as the Maastricht Treaty of the European Union and, less certainly (in the case of the current US administration), the Kyoto protocol on carbon dioxide emissions levels.

Held *et al.* (1999) argue that political power and activity has 'stretched' across the boundaries of the nation state and political issues have become globalized. For example, pollution, drugs, crime, terrorism and human rights are considered inappropriate issues to tackle on the basis of national political agendas. Social movement and pressure group activity is increasingly operating on a global scale, transnational companies are often larger and more powerful than state governments, and traditional areas of state responsibility, such as defence and economic management, are coordinated by intergovernmental bodies such as NATO and the WTO. For Held and colleagues, a system of global governance is emerging where nation states have been obliged to give up some of their sovereignty to larger political units. However, forms of transnational governance in developing countries may make less compromising demands on the sovereignty of a nation state, such as the OAU (Organization of African Unity; see Clapham 1985: 128). For Held (Held *et al.* 1999) and Lash and Urry (1994) the nation state is in some sort of decline, but Giddens (1990) argues that the nation state has always operated in contact with other states, and that, at certain historical junctures, this contact merely intensifies.

Global political organizations have been the focus of campaigning for radical environmental activists in the 1990s and early 2000s. The World Trade Organization emerged in January 1995 from the series of agreements thrashed out between 1990 and 1995 aiming to 'guarantee' worldwide 'free trade', known as the General Agreement on Tariffs and Trade, or GATT. The WTO in particular has been a focus for recent protests by 'anti-capitalist' demonstrators, an eclectic group of environmentalist, anarchist, anti-imperialist and Marxist activists, who received international media attention for their demonstrations at the venue of the WTO conferences in Seattle in 1999 and Genoa in 2001.

Francis Fukuyama (1992) and Samuel Huntington (1991) both agree we are moving towards a global liberal democratic political culture with a common commitment to market capitalism and individual rights. Ronald Inglehart (1990) argues a rather different case, that 'post-materialist' political culture in wealthy states of the globe brings them more closely into line with 'less developed' countries. Whether we all think in the same way about politics is difficult to assess, but some political issues do seem to be becoming global in policy-making terms. Such issues include human rights, which, although a contested issue, has become enshrined in a UN declaration.

The last few paragraphs and indeed this chapter so far, indicate that different people are talking about different things when they speak of 'globalization'. There is the planetary perspective adopted by the critics of globalization, the neo-liberal ideology and practice of globalization, and the subjective cultural understanding of what globalization might be. Globalization is so vague a concept, it is perhaps no wonder it is so contested and differently theorized and experienced. The rest of this chapter will examine issues of development and the environment which have become more 'global' in terms of public perception and policy-making, and the contested nature of globalization as a concept and as a process can be seen throughout.

The globalization of environmental problems

Ulrich Beck (1992), as we have already seen, has developed a highly influential perspective in environmental sociology. Beck's thesis of the risk society is that we live in a 'community of danger' due to the threat of hazardous side effects caused by modernization and mass prosperity. A sense of environmental risks and threats is an important part of Beck's thesis, for he argues that in late modern societies the relationship between society and the environment is being both contested and changed. We humans are no longer able to see ourselves as distinct and separate from 'nature' (1992: 81), so this key tenet of Westernized modernity is being challenged. In fact, modernity is being challenged as a set of social practices and institutions, and it is the side effects caused by modern structures which becomes the new framework within which we humans act.

I would agree with Beck that both richer Northern states and the poorer states usually of the South are now in a globalized situation of risk, but am not convinced that his model applies in the way he suggests. Key to Beck's notion of a 'risk society' is the idea that we have moved, or are moving, from a society in which some of us are under the threat of material deprivation, to one where we are under environmental threats caused by industrial production and consumption. 'Risk' is becoming a globalized phenomenon (Beck 1999). Yet in countries where malnutrition and starvation may be endemic, such material deprivation may structure our perception of risk immediately, and the politics of hunger is not, as Beck (1992) suggests, outdated. The distribution of 'risks' also varies, being unevenly concentrated in the poorer countries of the globe. In 'risk societies' the public is 'reflexive' in regard to risk, that is, they live with a permanent sense of hazard. Not all hazards are evenly experienced and similarly perceived however. As we saw in Chapter 3, the 'environmentalism of poor' does not only refer to the experience of risk of those living in poorer counties of the Southern Hemisphere, but also to those poorer people living in some of the most affluent regions of the globe (see Bullard 1993). As Bell notes, 'we cannot doubt that the rich are in a far better position to avoid the worst consequences of all these (environmental) threats' (1998: 194).

Despite such disparity, governments, media and environmental social movement organizations increasingly accept the notion of environmental issues as 'global' phenomena. The rest of this chapter accepts that there are such things as 'global environmental issues' to be examined, but

concentrates on their disproportional distribution in 'developing'
countries.

Global environmental issues

Issues such as water and air pollution, radiation, genetic modification of
foodstuffs, resource depletion, and depletion of biotic diversity have an
accentuated global impact in the twenty-first century. If we take pollution
as an example, both the global impact, and unequal distribution of
environmental problems may be illustrated. People in one country of the
globe may produce significant amounts of airborne pollution, but
countries elsewhere may bear the brunt in terms of its environmental
effects. In the well-known case of the devastation of Scandinavian forests
by acid rain, one of the key culprits is British electricity generation. A
particularly powerful example was the radiation cloud from the accident
at the Chernobyl nuclear power station in the then Soviet Union in 1987.
As people watched the progress of the cloud, and meteorologists
predicted its path across the globe, the planet may well have felt
'compressed'. Not all environmental problems can be seen as 'global',
however. Some forms of pollution are caused by, and have effects upon,
specific localities: a city, a beach, the banks of a minor river in one region.
This said, the interrelated quality of natural eco-systems means that
environmental problems often have impacts beyond the boundaries of
traditional nation states. The atmosphere and the climate are common
eco-systems for the planet, and although a minority of affluent states may
be responsible for much of the pollution that degrades them, the impact
of such actions is global.

David Goldblatt (1996) has argued that the environmental aspects of
globalization can be seen in two historical epochs: first, the expansion
of European trade from the sixteenth century to the Industrial Revolution,
second, the effects of such processes in the twentieth century. Agricultural
revolutions in eighteenth-century Europe quickened the pace of
environmental degradation in terms of loss of flora and fauna, and
pollution and exploitation of resources soared with the emergence of
the Industrial Revolution in the late eighteenth and nineteenth century.
However, the change from the nineteenth to twentieth centuries is that the
impact of environmental degradation has been felt in the last century only.

Global warming is probably the clearest example we have of a global
environmental problem, and, in the last 25 years, the monitoring of

climate change has led to the emergence of a consensus that there is a problem. Global warming derives both from industrial and agricultural processes. Industrial emissions from factories located in Northern countries and from car emissions, are disproportionately responsible, with a significant yet unacknowledged proportion of methane contributed by the digestive systems of huge agricultural animal populations from countries across the globe (for example, Brazilian and Argentinian cattle, see Rifkin 1994: 192–9). Its consequences are certainly global, with all parts of the world likely to expect some shift in climate, and therefore in the kinds of flora and fauna that are suited to their climatic conditions. Some of the problematic effects, such as the melting of ice sheets and consequent rise in sea levels, will affect certain low-lying countries of the globe and not those which are land-locked, but all forms of agriculture might suffer disruption.

Pollution has many forms and can affect the sea and waterways, land and air. Land pollution is often localized; humans pollute land where they bury industrial and consumer waste, or where chemical factories, for example, have been located. This might suggest that land pollution is a local and not a global issue, but, due to the international trade in commercial and industrial waste, and the practices of dumping waste in international waters and on other countries, there is a global dimension. In addition, the spread of Western industrial and agricultural practices can be seen to have spread land pollution across much of the globe. Industrial effluents, agricultural chemicals, and the discharge of human and animal sewage into rivers, pollute waterways. Westernization has led to pollution increasingly assuming global dimensions. While the dumping of nuclear waste at sea was halted in 1995, and sewage dumping has been restricted, at least for EU member states, since 1998, other problems remain, particularly those of oil leaks and spills from tankers. The pollution of the seas is global, as almost all sea water is connected and pollution spreads internationally. Such pollution is hard to monitor and its restriction difficult to enforce (Yearley 1996: 33–4). Air pollution is highly pervasive, and, as it is carried over considerable distances by prevailing winds, is global in character. Acid rain caused by such pollution comes largely from fuel-burning power stations and also results from domestic burning of fossil fuels. The case of polluting one's neighbour is common, as illustrated by British pollution affecting Scandinavian forests. Yearley (1996: 40–2) sees food as polluted through the globalized use of agricultural chemicals, genetically modified crops and the contamination of foodstuffs, such as the British case of BSE-infected cattle and the risk of disease 'leaping' the species barrier and causing new variant CJD in beef-eating humans (see Chapter 6).

The depletion of climatic ozone is a relatively recent problem. Its effects are concentrated at the poles, but, given the unknown processes of ozone chemical change and climate patterns, it may have a more global impact. The manufacture of CFCs has been an innovation that has spread globally, and its restriction has been global in policy terms. The globally compressing effect of CFC emissions is that CFC usage in one town or city could have an impact on the ozone a remote distance away, as CFCs are carried by wind and work their way up into the earth's atmosphere. Ironically, as Yearley remarks: 'The most extreme loss of ozone in fact occurs at the two poles, where presumably, there is less call for deodorant' (1996: 27).

Resource depletion can also be seen as an environmental problem of global proportions. We have a global energy market in fossil fuels that may well be in crisis in the next two decades if predictions of the depletion of reserves of petroleum, coal and natural gas are accurate. While new reserves are continuously found, these are not likely to last until the next century, and as demands, particularly from developing countries, increase yearly, this is probably an optimistic estimate. The alternatives are either uncertain to yield sufficient energy (solar and wind generation) or to have environmentally hazardous effects (radioactive by-products of nuclear power, and habitat destruction from mineral mining). Most serious is the possibility of localized shortages of water, as land-bound countries are using artesian sources (tapped by wells from underground reserves which accumulate very slowly) far more quickly than those reserves can be replenished. It should not be assumed that such reserves are used solely by land-locked and desertified counties of North Africa and the Middle East, for American agriculture and industry are also highly dependent on such underground supplies (Yearley 1996: 50).

Environmental social movement organizations have successfully defined the loss of species, biodiversity as a global environmental problem. Biodiversity refers to the amount of genetic diversity on the planet – how many species, and varieties within a species. This is particularly so in the case of large animals experiencing severe declines in populations, or potential extinction, such as pandas, tigers, rhinos and whales. The finger of 'blame' is often pointed at populations of Southern regions for hunting, and agricultural practices which limit the habitat of such species, conveniently forgetting that in much of Europe, for example, **mega-fauna** such as the wolf have been hunted to extinction. The spread of Western agricultural and industrial practices across the globe has resulted in rapid decreases in species diversity. This has been 'real' in the sense of declines in ocean biodiversity as a result of over-fishing, and 'symbolic' in the sense that '**charismatic mega-fauna**' such as pandas, tigers, orang-utans and whales have acquired global significance as symbols of environmental threats. Images have been used very effectively by environmental organizations, and representations such as that in Figure 5.1 often accompany requests for sponsorship and donations.

As we saw in Chapters 2 and 3, population growth has been a contentious issue, and I will not return to the details of the debate here. That certain regions might be seen by some as overpopulated by the poor, while those with declining populations are seen by others as over-consuming,

Figure 5.1 *Appealing for habitat preservation: 'charismatic mega-fauna'* (*Source*: David Lawson, WWF UK)

illustrates the point that global environmental hazards are embedded in the context of inequalities between rich and poor nations.

Environmental problems in the Third World

As most global environmental problems do not have uniformly global consequences, there is fierce debate over who is to 'blame' for causing environmental damage. Some environmental problems may not be 'global' in themselves but become global because they are dispersed through international trade and industrialization. Developing nations have resisted pressure to sign treaties on limiting industrial emissions, arguing that the wealthy Northern states were responsible for environmental damage in the first place, and should 'clean up' their own industrial production while accepting that developing countries should be able to modernize as they wish. Underdevelopment and environmental problems are closely linked, and environmental issues in developing countries are of concern to both indigenous populations and governments and international pressure groups and non-governmental organizations (NGOs). This is because:

> many environmental problems are at their most severe in the Third World (or the 'south'). It is there that the rain forests are being cut down, there that the toxic wastes are illegally dumped, there that most endangered species live and there that desertification threatens agriculture.
>
> (Yearley 1992: 149)

Underdevelopment, global trade and environmental problems are intertwined. For example, there are various social and environmental costs that stem from agricultural production in the Third World. Yearley (1992: 151–2; see also Latouche 1991) has noted that the plantation economy of many developing countries can only be understood in terms of Western demand for sugar, tea and coffee. The price of these products is established in the financial centres of the prosperous North, which means that developing countries make limited profit on their sale. Single-crop plantations use valuable land that could be used for crops for the domestic market. Plantations also have environmental costs, such as susceptibility to disease, which in turn promotes the use of chemical pesticides and reduction of soil fertility then necessitating the use of chemical fertilizers.

In addition, the dependency of Third World countries makes them susceptible to environmentally harmful industrialization due to 'regulation flight' (Yearley 1992: 158). Developing countries tend to attract investment from transnational corporations due to lower wages, weaker trade unions and health and safety legislation, and less stringent controls on pollution. Some Third World countries see pollution as a necessary price to be paid for modernization, while companies with businesses which pollute seek out new investment opportunities in countries where environmental regulations are weak. This is particularly the case with the chemical and mineral-processing industries. For their part, transnational corporations are often able to pollute the natural environment with impunity, and to subject their workers to dangerous substances and practices as health and safety regulations are weak.

Agribusiness TNCs, such as Nestlé, Heinz and Del Monte, are particularly attracted to Third World countries because of their often-favourable climates for growing crops such as exotic and soft fruits. Susan George (1988) has shown that such companies use the most productive land in such countries to grow cash crops for export, leaving those countries needing to import food and undermining their ability to be independent in food production. Plantation farming for export crops reduces soil fertility, necessitating use of chemicals that further degrade the soil. George argues that the cash-crop-producing TNC's employ few plantation labourers, and that governments of countries such as Brazil in the1970s and 1980s were prevented from undertaking land reform by the plantation-dominated agricultural economy. Consequently, the unemployed poor were encouraged to eke out a living cultivating the unsuitable land of the Amazonian rainforest, thus depleting the forests and coming into conflict with indigenous animal and human inhabitants. So international relations of economic dependency may result in human unemployment, poverty and displacement, and environmental problems of depletion of rainforest fauna and flora, soil infertility and soil and water pollution (Merchant 1992).

As the nations and regions of the globe come together in global political and economic forums, a common concern with the environment does not necessarily lead to shared interests. The Third World and the industrialized countries experience differing impacts and forms of environmental problems, and have different agendas. The developed world tends to prioritize issues such as global warming, ozone depletion, deforestation and habitat conservation, whereas developing countries are more interested in the economic relations between rich and poor countries, and the deleterious impact of Northern industrial practices, trade agreements and consumer culture on the global environment.

Global political institutions and the environment

Yearley (1996) contends that governments and environmental movements are working both to define environmental problems as 'global' in nature, and requiring management beyond the boundaries of the nation state. In some cases, such as the trade in rare species and exotic wildlife, there have been internationally binding restrictions for over a century. But the 'greening' of political responses to environmental problems, and the international character of such responses, is largely a recent development.

Since 1945, much of the development of international environmental policy has taken place under the auspices of the United Nations through its agencies such as UNESCO (UN Educational, Scientific and Cultural Organization) and the WHO (World Health Organization). Despite the inhospitable climate of the Cold War, some environmental protection initiatives were successfully agreed, although their implementation was less successful. At this point, many Third World countries were still colonies of European states and were not represented nor acknowledged to have any other interests than those of their colonizing governments. The convention regulating international whaling was signed in 1946, and some regulations on nuclear and other toxic waste disposal and some forms of habitat protection were agreed in the 1950s and 1960s. The first international initiative of significance was the 1972 Stockholm Intergovernmental Conference on the Human Environment, sponsored by the UN Environment Programme (UNEP). As a result of the conference, the developing countries campaigned to have the UNEP secretariat located in a developing nation and in this they were successful – the secretariat was located in Nairobi, Kenya (Miller 1995: 8). National governments, environmental organizations and a range of non-governmental organizations (NGOs) were represented. As part of the preparation for the conference, a report was produced and published for public consumption, based on the advice of expert advisers on the then current state of the global environment. The authors assumed that environmental risks were commonly experienced and that there was a global interest generating a rational 'loyalty' to the planet (Ward and Dubos 1972: 32, see Box 5.2).

Over a decade later, the UN Commission on Environment and Development produced The Bruntland Report of 1987, *Our Common Future*. This report also argued in strong terms that states need to cooperate in radical action to combat and avert environmental hazard, that this was in their self-interest, and that the 'interest' of all states was the

---

Box 5.2

The UN: a global and anthropocentric environmentalism?

Excerpts from *Only One Earth* (1972) a Report Commissioned by the United Nations, and written by Barbara Ward and Renée Dubos:

> Now that Mankind is in the process of completing the colonization of the planet, learning to manage it intelligently is an urgent imperative. Man must accept responsibility for the stewardship of the earth . . . the charge of the U.N. to the Conference was clearly to define what should be done to maintain the earth as a place suitable for human life, not only now, but also for future generations.
>
> (1972: 25)

> There was general agreement among the experts that environmental problems are becoming increasingly world wide and therefore demand a global approach.
>
> (1972: 28)

> We are indeed travellers bound to the earth's crust, drawing life from the air and water of its thin, fragile envelope, using and reusing its very limited supply of natural resources. . . . The fundamental task of the U.N. conference on the Human Environment is to formulate the problems inherent in the limitations of spaceship earth, and to devise patterns of collective behaviour compatible with the continued flowering of civilizations. . . . The emotional attachment to our prized diversity need not interfere with our attempts to develop the global state of mind which will generate a rational loyalty to the planet as a whole. As we enter the global phase of human evolution it becomes obvious that each man [sic] has two countries, his own and Planet Earth.
>
> (1972: 31–2)

---

same. Over three decades, these reports of 1972, 1987 and 1992 have assumed that we are part of a global community, that our interests are similar, and that actions by nation states and international organizations can apply across the globe. Such an appeal to self-interest is mistaken, however, argues Yearley (1996), because while nation states do act overwhelmingly in terms of their own self-interest, their interests are very different.

Some governments may favour high environmental standards not because they are in any way concerned about the environment, but because high environmental standards favour the operation of their industries. The latter already have high standards and risk being undercut by industries in countries that do not. 'First World' countries of the Northern Hemisphere have been accused by environmentalists in developing countries (Shiva 1993) of using the concept of global interest as a cover for their own self- or regional interest. In addition, environmental policies may serve the interests of transnational corporations. The Vienna Protocol of 1985 and the more stringent Montreal Protocol of 1992, banned the use of ozone-depleting CFCs. The American-based TNC Du Pont supported the later initiative because they had developed dismantling technologies and CFC alternatives. For countries like Malaysia, that had only just developed the technology to produce and sell CFC-dependent products, an international ban was most undesirable (Yearley 1996). The relocation of 'dirty' Western-based companies to developing societies illustrates that environmental problems are often caused by social factors of poverty, debt to international agencies and the disparities of the international trade system.

Since the 1970s, governments and international institutions have taken a range of 'global' policy decisions and commitments, some of which have been relatively uncontroversial in terms of the divided interests between 'First' and 'Third' Worlds. The London Dumping Convention (1972), the MARPOL convention (1978) on dumping at sea and the UN Law of the Sea (1982), for example, covered the control of marine pollution. There have been a series of wildlife protection measures with agreements on polar bears, whales, seals, trade in rare species, migratory birds and some mammals, and preservation of the Antarctic region. Conventions were signed on the transport and disposal of hazardous waste (Basel Convention, 1989). In the European Union, NAFTA, the Nordic Council and other international political institutions, environmental policy has become a key subject for both dispute and debate, and forms of international cooperation (Held *et al.* 1999: 387–9).

Highly controversial in recent times, however, have been the debates on world trade, and this has raised important questions for environmental policy as well. After the setting up of the UN in 1945, financial markets were managed by the International Monetary Fund. Since 1995, these markets have been managed in a decentralized way through the World Trade Organization, which emerged from the annual talks of the GATT, the General Agreement on Tariffs and Trade, which had until 1995 been coordinated by the G7, the leading world capitalist states. Since its inception, developing countries have voiced severe criticisms of the operation of the WTO, for its policies are seen to enhance the interests of the already prosperous nations of the globe. Wolfgang Sachs (1993) is strongly critical of the role of the UN, arguing that the discourse on 'development' since the 1980s has operated in an uncritical pro-growth Western paradigm. UN-sponsored 'development', in his view, has a hidden agenda of Westernization. This may be so, but is likely to be caused, as Held and colleagues (1999) argue, by the globalization of economic production which has enhanced the power of corporate capital, and increased its influence in supranational political institutions. The anti-globalization protests which have achieved media prominence through protests at the 1999 World Trade Organization conference in Seattle, and 2001 conference in Genoa, unsurprisingly perhaps, campaign on an environmentalist as well as anti-capitalist platform.

In a perhaps belated follow-up to the Stockholm conference of 1972, came the First 'Earth Summit', an intergovernmental conference held in Rio de Janeiro in 1992. Conducted under UN auspices, and with almost every member state represented, Rio sought to establish the most far-reaching set of global agreements and targets on environmental protection. It included conventions on the rainforest, climate change and greenhouse gases, and biodiversity. The impact of Rio was significant in that it placed environmental issues on the agenda of global institutions such as the IMF and World Bank, and would determine that these should be within the remit of the WTO.

One of the undercurrents of the 1992 Rio conference however, was the dissatisfaction of governments of poorer countries with the Western domination of the issues debated and the decisions made. In examining the impact of Third World countries on global environmental policy-making, Marion Miller (1995) argues that, despite the dominance of Northern nations, the global impact of environmental problems does give developing countries some leverage. Policies for sustainable development are becoming more widely discussed, and the World Bank, post-Rio, reorganized its environmental management processes and created a vice-presidency in Sustainable Environmental Development (Miller 1995: 28). However, in recent history, the role of the World Bank and IMF, institutions which have supplied some of the economic advice for international environmental protection agreements, have also had an important role in diffusing globally Western models of economic development (Sachs 1993). Many developing countries are sceptical of the apparent change in the orientation of the World Bank. This is particularly so given the increasing influence of the American-based TNCs in a globalized economy (Shiva 1998). It is to the role of the WTO and the American-based TNCs that we will now turn in considering the management of biotechnology, a relatively 'new' issue of environmental concern which has aroused the hostility of environmentalists in both rich and poor regions of the globe.

Transnational corporations and biotechnology

Many environmental campaigners see transnational corporations as the most significant players in causing environmental hazards. In large part, this perception has led to the globalization of protest and a focus on the World Trade Organization in particular, as it is seen to be dominated by the interests of the largest American-based transnational corporations.

Sharon Beder (1997) examines the power of the transnational companies to influence public opinion through the media, the academic and industrial science establishment and corporate advertising. She argues that the global economy does not operate the free trade system that the WTO is supposedly there to ensure. Rather, free trade means in effect, monopoly by a small number of transnationals all based in the USA: Du Pont, General Electric, Bristol Myers, General Motors, Monsanto, Rockwell, Warner, Johnson & Johnson, IBM, Hewlett Packard, Marck and Pfizer.

For Vandana Shiva, the free trade system is of particular concern to environmentalists because it involves the imposition of monopolies of genetically modified products. The final treaty of the GATT is the TRIPS (Trade Related Intellectual Property Rights) agreement. The globalization of patenting and intellectual property rights through TNCs' use of TRIPS agreements causes a loss of species diversity, ecological destruction and the imposition of **monoculture** in society and the environment (1998: 82). For example, at the World Food Summit in 1996, Monsanto claimed that its soya bean was distinctive due to genetic engineering, and was able to patent the seed. Under the GATT regulations, farmers in India can buy the Monsanto seed, plant and harvest it, but not save the seed for replanting the following year – they must buy more from the original source. Shiva claims that this is a disruption of traditional agricultural practices based on the free exchange of seed, that it limits the biodiversity of crops and constitutes a 'theft' of the genetic diversity of the Third

World. The spread of the use of 'superseed', such as the Monsanto bean, could result in the proliferation of super-weeds and superpests, which will evolve in response to more robust plants and the pesticides Monsanto sell to accompany their beans. Genetic modification of crops and, ultimately, the management of agricultural production, according to Shiva, will not be the prerogative of communities or fall within the jurisdiction of nation states. Rather, it will reflect the interests of the TNCs as the main players in a globalized polity and economy, and be at the expense of Third World societies and their environments.

The states of the South have not accepted this development without contestation however. Indian domestic legislation, the Indian Patent Act, had strong licensing clauses to uphold 'the public interest' in safe food. When, in 1995, the Indian government, in line with GATT regulations, introduced a Patent (Amendment) Bill to allow product patenting in foodstuffs, it was rejected by the Indian Parliament. The Indian Farmers' Movement, which formed in 1993, argued that, for the Indian farmers, to produce, exchange and sell seed is part of their birthright as citizens of an independent Indian state (see Shiva 1993). In 1996, they adopted a policy of 'seed *satyagraha*', a method of non-violent direct protest action following the methods of M.K. Gandhi, which involved the disruption of the seed trials. In May 1998, Monsanto applied to the WTO, urging it to compel India (by threat of expulsion) to implement the TRIPS agreement and ensure that GM food trials could proceed. The Indian government refused pressure from the WTO and, in October 1999, Monsanto withdrew its complaint. A nation state and grassroots social movement called a halt to the application of biotechnology for two and a half years. In 2002 however, Monsanto has the 'right' to sell its seed, or, as Shiva sees it, to 'force hazardous products on people' on grounds of free trade (1998: 43).

Conclusion

While environmental problems, issues, risks and hazards are being treated by the mass media, social movement organizations, governments and international institutions as global in character, their 'global' nature is questionable. Some problems can be seen as global in the scope of their effects, but in most cases, problems are distributed, experienced and caused unevenly. The differing environmental priorities of societies in different regions of the globe reflects the main theme of this chapter – that looking at social, economic and political practices helps make sense of current environmental problems and helps in understanding their complexity. In addition, environmental problems may be caused by social practices associated with the development of modern industrial and consumer societies. Globalization itself is a highly contested term, with widely differing perspectives on the nature of globalization, the processes of change, its historical development and long-term future. While international organizations such as the UN have attempted to discuss initiatives and develop global policy programmes with respect to the environment, again, different kinds of social, economic and political interests can conflict in such attempts.

While environmental problems may affect 'us all', activists, politicians and policy-makers need to be sensitive to the differing perceptions of such problems and issues in order to develop policies which genuinely attempt to reflect the interests of the diverse societies of the globe.

Key points

- The concept of globalization is popular, yet contested, within the social sciences. It assumes that different societies are increasingly linked through trade, political organizations and cultural exchange. Developments in travel and, in particular, communications media have led, some argue, to a global awareness.

- Globalization theory has its roots in theories of modernization and development. Modernization theory attempted to apply models of social change developed in the context of eighteenth- and nineteenth-century Europe to the 'developing' world. Neo-Marxists have used dependency and world systems theories to explain why developing countries were not developing along similar lines to the countries of Europe because of exploitative trade relations.

- Different perspectives on globalization cut across the left/right political spectrum. Some see globalization as a new epoch in the organization of human society, representing the demise of the nation state and the triumph of liberal democracy and market capitalism. Sceptics argue that political and economic relations have not intensified in the twentieth century, and some contend that globalization is a form of economic and cultural imperialism. Others argue that there is a process of increased interconnection between societies, but that the nature of this process is complex and unclear.

- Globalization can be used to explain certain changes in contemporary society such as the de-nationalization of large firms, the spread of certain cultural images, and the internationalization of certain kinds of policy-making. Environmental policy is a case in point.

● Policy-makers, the media and environmental social movement organizations are increasingly representing environmental issues and problems as global. These issues include global warming, desertification, land, water and air pollution, ozone depletion and biodiversity. However, the causes and consequences of environmental risks are spread unevenly across the regions of the globe. The social and economic circumstances of the poorer counties mean that the effects of environmental problems are particularly severe for them.

● Since the 1970s, there have been a number of 'global' initiatives in environmental policy, often under the auspices of the United Nations. Some of these measures have been controversial, as they are often seen by governments of developing countries to reflect the interests of wealthy Northern states of the globe.

● Particularly controversial has been the role of the World Trade Organization in relation to the transnational corporations. TNCs can be seen as important players in global policy-making, and have been the subject of environmental concern in their attempts to globalize chemical-based agriculture.

Further reading

There is a wide range of material on globalization at present. Perhaps the most comprehensive text, which covers a range of disciplines within the social sciences and many issues, including a section on the environment, is David Held, Anthony McGrew, David Goldblatt and Jonathan Peraton, *Global Transformations: Politics, Economics, Culture* (Cambridge: Polity, 1999), and its partner, the *Global Transformations Reader* (Cambridge: Polity, 2000). A useful introduction to the concept is Malcolm Waters's *Globalization* (London: Routledge, 1995), particularly Chapters 1, 4 and 5.

On modernization and dependency theory, Vicky Randall and Colin Theobald give a succinct account of the debates in *Political Change and Underdevelopment* (London: Macmillan, 1985). World systems theory is comprehensively presented by Leslie Sklair's *The Sociology of the Global System* (Hemel Hempstead: Harvester, 1991), and for a brief and specific application to the environment see his 'Global Sociology and the Global System', in Michael Redclift and Ted Benton (eds), *Social Theory and the Global Environment* (London: Routledge, 1994). For more detailed accounts of different positions within the globalization debate, see Alex Callinicos' edited collection of neo-Marxist sceptical accounts in *Marxism and the New Imperialism* (London: Bookmarks, 1994), Anthony Giddens's *The Consequences of Modernity* (Cambridge: Polity, 1990) and Martin Albrow's *The Global Age* (Cambridge: Polity, 1996).

On globalization and the environment, with a detailed examination of global environmental issues, the most comprehensive text is Steven Yearley's *Sociology, Environmentalism, Globalization: Reinventing the Globe* (London: Sage, 1996). The chapter on 'Development and the Environment' in his text *The Green Case* (London: Routledge, 1992) is also useful.

More material on environment and development can be found in Michael Redclift's paper 'Sustainable Development and Global Environmental Change: Implications of a Changing Agenda', *Global Environmental Change* 2, 1 (March 1992): 32–42. On policy-making, see Marion Miller, *The Third World in Global Environmental Politics* (Buckingham: Open University Press, 1995). For a developing world perspective, see Serge Latouche, *In the Wake of the Affluent Society* (London: Zed, 1991). For a lighthearted and lively account, including a marvellous description of President George Bush (Senior) accidentally clapping Fidel Castro's speech denouncing environment-ravaging capitalism at the 1992 Rio Summit, see Chapter 13 of Susan Calvert and Peter Calvert's *Politics and Society in the Third World* (London: Prentice Hall, 1996). A provocative account of biotechnology can be found in Vandana Shiva's *Biopiracy* (Dartington: Green Books, 1998).

Discussion questions

1 In what ways, in your experience, is the world 'getting smaller'?

2 Are different perspectives and debates on globalization any improvement on those on modernization and dependency?

3 Should the poorer countries of the South be able to develop as they wish or 'must' they develop in a 'sustainable' way? Who can and should decide?

4 What does the economic power and political influence of transnational corporations mean for environmental policy-makers and social movements?

Society, 'culture' and 'nature' – human relations with animals

This chapter will:

- examine the ways human–animal relations have changed historically, using the example of modernization in Europe
- compare and contrast different theorizations of contemporary human–animal relations
- critically examine the debates and social movements within the politics of animal rights
- discuss three different examples of arenas where modern Western society interacts with animals: the production and consumption of animals as food, 'pet' keeping, and the use of animals as a means of human entertainment

Introduction

This chapter examines how sociologists and other social scientists have viewed the relationship between human society and non-human animals. The last three chapters have been organized around various themes in environment–society relations. Chapter 3 is concerned with politics, Chapters 4 and 5 with space, place and social inequality. The theme addressed in this chapter will be 'species'. Why then, animals, particularly mammals, and not plants? First, our relations with higher mammals are different from those with plants and landscapes, and the latter has been addressed, to some degree, in Chapter 4. Second, the questions raised by the issue enable us to draw on the insights from the sociology of 'everyday life', as well as giving us an example to illustrate some lively sociological disputes. The issues of Chapters 3–6 are not, of course, entirely distinct. Human relations with animals differ vastly cross-culturally and historically, and animals have a different material and symbolic existence in agricultural and industrial spheres of economic production and in urban and rural spaces. Human–animal relations is a political issue and animal rights is itself a distinct strand of green social movement activity. Issues upon which animal welfare organizations have campaigned involve colonial, postcolonial and globalizing processes: the whaling industry, the trade in ivory, fur and feathers, the 'preservation' of wilderness and habitat, for example.

The term 'animal' is problematic. Humans are, of course, animals, and some theorists have argued that the distinction between humans and 'animals' is a fiction – a social construct that subordinates animals and privileges human beings. Although the term 'animal' may be contested, due to its prevalence in the literature and its normative everyday usage, this chapter uses 'animal' to refer collectively to the whole variety of non-human animal species.

Reflecting this notion that 'animals' are, at least partially, a product of human society, the chapter begins by looking at how social ideas about animals have altered historically, and, in particular, how the changes which accompanied modernization have affected human relations with animals. While the literature in some other disciplines (anthropology, for example) takes account of the various ways different cultures conceptualize animals, we concentrate here on changes in Western thinking from the sixteenth century onwards as an example of how social conceptions of animals are dynamic and shift across time. Most of the sociological literature on human relations with animals is of recent origin, and we look at a number of ways in which contemporary sociologists have theorized such relations. The chapter also examines the concept of 'animal rights', which has increasingly become part of political debates in Western societies, and has influenced popular thinking on the appropriate relations between humans and other animal species.

The final sections of the chapter look at different aspects of contemporary relations between human societies and animals. In everyday life, people interact with animals continually. Most of us will eat animals and animal products, wear animal products, and many of us share our homes with animals. Even if we choose not to watch them, in parts of Europe and the United States, we are aware of the many television programmes which feature animals: wildlife documentaries, 'vets and pets' docudramas, cartoons and children's films. A significant minority of us will chose to view animals 'live' by visiting zoos, aquaria, city farms, safari parks and other entertainment venues which display animals for the public. Some of us may engage in a 'sport' which involves animals, such as hunting, shooting or fishing. The chapter looks at how these practices of eating, observing, hunting and keeping animals have changed in recent decades, and considers how some sociologists have interpreted these changes.

Historical change and human–animal relations

In a comprehensive account of human–animal relations in Western societies, Adrian Franklin (1999) argues that from the nineteenth to the end of the twentieth century, we have seen a dramatic alteration in human relations with animals. He suggests that people now spend more time with animals and do things involving animals more often than they did a century ago, and posits that the quality of our relationships with animals has changed significantly. Keith Thomas (1983), analysing changing attitudes towards animals in early modern England from the sixteenth to the end of the nineteenth century, likewise sees some dramatic shifts in both popular perception and in the material treatment of animals. Thus relations between that part of 'nature' which biologically is closest to us, other higher animals, shifts over time. In addition, both Thomas and Franklin argue that the changing perceptions of animals tell us much about social relations between humans, and that, historically, nature and 'society' have not been entirely separate entities, but are interwoven – animals are part of human cultures.

According to Thomas (1983), in Tudor England, most people lived a predominantly rural existence in close proximity to domestic animals, with which, in winter, they were often compelled to share accommodation. Animals were crucial to human ways of life, providing food, transport and labour power. Human relations towards other animals were, in Thomas's view, characterized by both contingency and 'anthropocentricity' (human-centredness), as, according to Christian orthodoxy, humans had absolute rights to use animals as they saw fit. This does rather homogenize Christian thought. The ascetic Calvinism of the sixteenth century, which Weber saw as so important in generating the cultural conditions for the development of capitalism, was undoubtedly extremely hostile to 'nature' in human or animal form! However, John Passmore (1980) argues that Christian thought is also characterized by a 'stewardship tradition', which is ancient Greek in origin, and sees non-human nature as a divine creation over which humans must be responsible stewards. Passmore acknowledges that the anthropocentric discourse came to dominate the Christian worldview, and I think Thomas's emphasis on human domination in the Christian tradition is uncontroversial.

Attitudes to animals changed with the onset of the processes of economic, social and political modernization in England in the seventeenth century. The emergence of scientific knowledge undermined the earlier theological view of the world and, by the nineteenth century, had defined humans as 'mere' animals among many (Thomas 1983: 166–7), albeit at the pinnacle of the evolutionary ladder. Society grew less dependent on animal power with the advent of mechanization, and the difference between humans and animals increased with the spread of urbanization. Thomas also suggests that people increasingly sentimentalized animals, as they decreased in utilitarian significance, and, in the eighteenth century, the practice of pet-keeping grew in urban areas. The sentimentalization and proximity to companion animals encouraged the notion that animals were individuals with personalities, and Thomas sees this as linked to the formation of urban-based movements for animal welfare and debates about 'animal rights' at the turn of the nineteenth century (Thomas 1983: 119). Thus, for Thomas, urbanization forced a separation from and then reconceptualization of human relations with animals. This was combined with a less utilitarian view resulting from the changes associated with industrialization, and the boundaries placed on anthropocentrism with the development of scientific knowledge.

Franklin (1999) argues that Thomas gives too much weight to the effects of urbanization, that the division between rural and urban society was not so clear-cut, and that animals continued to have utilitarian uses as transportation in urban areas. Keith Tester (1991: 71) considers that Thomas places excessive emphasis on 'pet-keeping' in altering sensibilities towards animals, arguing that the practice was not confined to urban dwellers, and that, in any case, the practice of pet-keeping cannot explain the ideological origins of 'sentimental' attitudes towards animals. In addition, Andrew Johnson (1995: 170) cites the historical research of Kellert and Westervelt (1982) which indicates that, during the period in question, it was not that concern with animal welfare and sentimentality increased, but rather that, with advancing modernization, a more detached view of animals came to predominate.

This view is echoed by some eco-feminists who have examined historical links between formations of gender and human relations with animals. Carolyn Merchant (1980: 3–5) suggests that modern Western scientific culture established the notion of a hierarchy of species and legitimated human domination over animals. The pre-modern European worldview had a generally benign view of nature, wherein animals were 'ensouled beings', but with modernization they were reconceptualized as machines. Similarly, Vandana Shiva argues that in pre-colonial India, under the influence of Hindu philosophy, animals were seen as having spirit. It was the colonial spread of scientific rationalism that defined non-human animals, along with the rest of 'nature', as inert objects. Brian Eastlea's (1980) work indicates that Merchant romanticizes the worldview of Renaissance Europe, because, while this was organic, it was also very hierarchical, and humans were higher than and dominant over, all other kinds of animals (according to the Christian doctrine of the 'Great Chain of Being'). Thus modernity may have reinforced a hierarchy between humans and other animals that was already in existence. There are interesting parallels however, between the nineteenth- and early twentieth-century European fascination with the scientific classification of classes, types and genealogies, and those of species. As Donna Haraway (1989) has noted, discourses of racism, gender relations and class, as well as the social construction of 'nature', permeate scientific writings and theories about animals, particularly primates. The classification of 'rare' species from the Southern Hemisphere was part of the cultural process of colonialism and the constitution of Eurocentric notions of civilized and uncivilized places and spaces, including their human and animal populations.

Tester (1991) concentrates on the imposition of social relationships on human relations with other animals. He deploys the perspective of Norbert Elias (1994) in contending that the development of anti-cruelty legislation was part of the 'civilizing process'. By preventing cruelty to animals, the state also prevented certain forms of working-class leisure involving violence and alcohol, so animal welfare legislation was part of a move to discipline the working class (Tester 1991: 68–88). Tester argues that human–animal relations are based around two competing 'discourses'. The 'Demand for Difference' emphasized the differences between humans and animals, and the civilized humanity of the upper and middle classes was 'a stick to beat social unruliness and "beastliness"' of the lower social orders (1991: 88–9). This perspective was championed by middle-class social reformers. In contradistinction, the 'Demand for Similitude' stressed the similarities between humans and other (higher) animals, and encouraged a more 'natural' human existence (1991: 89). Such an attitude was advocated by political radicals and romantic poets and represents the beginnings of contemporary demands for animal rights.

So, while human relations with animals have altered in Western societies over the last 300 years, within each historical period there have been different and competing conceptions of how humans can, and, as we will see later, how they *should*, relate to other animals.

Theorizing contemporary human–animal relations

We discussed sociological approaches in the first chapter, and examined their general application to the environment. I want to consider three different accounts of human–animal relations here, which are representative of extreme relativism, more critical social constructionism and critical realism. First, I will return to Keith Tester's (1991) postmodern account of human–animal relations; second, Ted Benton (1993) provides us with a realist analysis; third, Adrian Franklin (1999) has a broadly social constructionist account which takes some aspects of the realist critique on board.

Social constructionism

As we saw in Chapter 1, Tester (1991) denies that 'nature' has any causal powers or objective properties (i.e. properties which are independent of that which human beings ascribe). This means that the environment is essentially a product of society, and, as far as animals are concerned, they do not have a nature or being in themselves; they only have whatever being we humans decide to give them. So:

> A fish is only a fish if it is socially classified as one, and that classification is only concerned with fish to the extent that scaly things living in the sea help society define itself. After all, the very word 'fish' is a product of the imposition of socially produced categories on nature . . . animals are . . . a blank paper which can be inscribed with any message, any symbolic meaning that the social wishes.
>
> (Tester 1991: 46)

Ontologically (i.e. in terms of what we think about the nature of things existing in the world), animals only exist in the ways humans imagine them. Epistemologically (i.e. in terms of how we know what we know about the world), Tester is also a strong social constructionist, for all we can know about animals depends on social interpretations. As Luke Martell (1994) has pointed out, this over-emphasizes the power of 'society' in determining the world. While how we classify animals such as fish inevitably depends upon what kinds of classifications we humans have developed, what we know about animals is also dependent on certain properties which pertain to animals. 'Fish' are not arbitrarily classified, but have certain objective properties (such as living in water and having scales) which we use to distinguish them from certain other sorts of creatures. Andrew Collier (1994) suggests that, if we accept Tester's characterization of animals as being whatever we would want them to be, then solving marine pollution would be easy: 'we could reclassify lumps of untreated sewage as "fish"'! (Collier 1994: 89).

For Tester, how we think about animals does not tell us about the ontological condition of animals, but about ourselves. So animal rights 'is not a morality founded on the reality of animals, it is a morality about what it is to be an individual human who lives a social life' (Tester 1991: 16). Animal rights has nothing to do with any concern for sufferings humans may inflict upon animals, but is about humans making themselves feel 'good' as moral agents arguing for those who cannot argue for themselves (1991: 78). Some element of realism, therefore, seems key to arguments for animal rights or welfare, for if animals cannot be seen as independent beings, which are able to feel or flourish, then they cannot be ill treated.

Donna Haraway also has a strong constructionist conceptualization of animals. With reference to scientific studies of primates for example, she contends that studies of animal behaviour tell us nothing about the animals themselves, but do tell us about the social locations and political opinions of the people who undertook the research (Haraway 1989). Elsewhere, she describes animals as 'blank paper' for human inscriptions (Haraway 1991: 6) and contends that modern societies are increasingly populated by '**cyborgs**', that is, that the boundaries of distinction between humans, nature (other animals and plant life) and machines are becoming increasingly blurred and permeable. More recently, she has examined contemporary developments in biotechnology and argues that we are fast moving towards a situation in which the 'natural' world becomes 'fully artifactual' (Haraway 1997: 108). For example, she sees the 'OncoMouse', the world's first patented mammal, created for vivisection purposes with cancer-bearing genes, as an interesting cyborg, and does not consider the treatment of mice in medical research to be problematic. Like Tester, she places too much emphasis on the social, and, to use Martell's (1994: 173) term, both Tester and Haraway 'sociologize away' any concern for animals. While Haraway acknowledges that humans and animals physically interact, her cultural analysis avoids looking at the complexity of the relationship by focusing only on the representation of animals. In my view, different material forms of human–animal relations need to be accounted for, and critical realism can provide a less abstract, more nuanced and more appropriately political analysis.

Critical realism

Realism, while critical of some of the bias present in scientific theories of animal ethology (the study of animal behaviour), accepts that, being independent biological beings with their own physiological and psychological needs, animals exist within and also beyond, human understandings of them.

Benton (1993) has sought to develop a theory of human–animal relations which is based on aspects of animal rights theory, but develops a distinctly eco-socialist perspective. He incorporates certain ecological insights, such as the concept of natural limits on resources, with socialist and feminist theories of equality and rights in order to apply them to the specific case of animals. Whereas Tester sees animal rights emerging in the nineteenth century as a discourse about human morality, Benton argues that it emerged as 'one response to an increasingly sharp contradiction of urban-industrial capitalism' (1993: 60).

Benton (1993) draws upon biological and anthropological studies of the 'social life of animals' in order to argue that many species have overlapping forms of 'species life' with humans. He argues for a **naturalistic** understanding of human society in which humans are seen to be both biologically embodied, with certain animal needs (food, sex) and socially and ecologically embedded. He draws on materialist eco-feminism here, in arguing that current social arrangements problematically ignore human need, vulnerability and dependency in gendered ways (1993: 103). Higher animals possess complex forms of social coordination (1993: 36), and Benton challenges the presumption of human separateness from 'other' animals. He argues that we should think about 'differentiations' rather than differences between animal species (1993: 45–57). Differentiations of species, and different social, economic and ecological contexts give rise to different categories of human–animal relationship, which I have labelled rather differently below:

- *animals as a labour force*: from the unsophisticated carrying and pulling of heavy objects (donkeys) to the sophisticated (guide dogs);
- *animals as food*: meat, milk, eggs, and to satisfy other needs such as clothing and shelter. Other human 'needs' such as medical vivisection;
- *animals as entertainment*: including fights of domestic and wild animals (bulls, badgers, dogs), hunting wild animals, fishing, shooting;
- *animals as edification*: in the media, particularly documentaries;

- *commercial exploitation of animals*: involving the extreme imposition of commercial requirements, for example, factory farming, commercial vivisection;
- *animals as household companions* or 'pets';
- *animals as symbols*: representing certain human qualities, for religious purposes;
- *'wild' animals*: existing outside incorporation into human social practices, or in conditions of limited incorporation.

(after Benton 1993: 62–8)

I find it difficult to accept his view that, while hunting may involve cruelty, circuses do not, because animals are 'not killed', and that in zoos 'animals are simply observed and enjoyed'. Simple observation can be intensely cruel if an animal is unable to move, is not fed properly, or if it is lonely or bored. Benton is careful to make the point that different conditions apply for different species, and argues that, in different forms of human–animal relations, different ethical considerations apply. Although I might have disagreements with the detail, the areas mapped out provide us with a complex picture of human–animal relations.

Benton argues that humans and animals stand in social relationships to each other, that animals are constitutive of human societies and that these relationships are incredibly varied across time and cultural space. These relationships are fundamental to the structuring of human societies: we are socially interdependent with animals and also ecologically interdependent (1993: 68–9). In arguing that animals are creatures with different constitutions, different species of which are in different social relationships with humans, Benton is questioning the divide between humanity and animality to which theorists of 'animals rights' are opposed. He is also advocating a fundamentally different position than the idea that animal relationships are ideological constructions with purely ideological functions, a view that denies the material co-dependency of humans and animals.

Benton's contribution to the debate on animal rights is that certain aspects of human rights carry across the species boundary very badly (1993: 210). He suggests that the difficulty with the rights discourse is its inability to take account of the prevailing social structures and relations of certain places at certain historical junctures. He rightly argues, in my view, that 'under prevailing patterns of animal use and abuse' rights are not likely to do much to alleviate animal suffering. Because he sees animals as in and of human societies, as 'co-evolving' (1993: 211) with them, we need fundamental changes in human social practices before we will see any shift in the treatment of animals. For example, he argues that the abolition of factory farming is a moral imperative, but that in order for this to be achieved, the economic relations of capitalist agriculture will need to be changed, as will the social organization of farming.

I think Benton's most significant sociological contribution to the debate on human–animal relations is the notion of animals as having species-specific relations with humanity, and as embedded in social relations. In drawing upon biological and anthropological insights, he provides us with a sociological account that is not reductionist and grapples with the insights of other disciplines. In Chapter 1, I suggested that my preferred environmental sociology would be one in which sociologists are not wary of drawing on various disciplinary knowledges in order to capture the complexity of social life. In terms of the sociological literature on human–animal relations I find Benton's work the most challenging and useful.

A softer constructionism

Franklin (1999) also calls the division between humans and animals into question, but in a very different way. He develops the case made by David Cooper (1993) that human–animal relations have shifted in recent times and have become increasingly 'sentimental'. Modern cultures involve closer emotional links between animals and humans, and the categorical boundary between human and other animal species has been challenged with 'postmodernization'.

The social causes of such shifts in human–animal relations are 'ontological insecurity', risk and misanthropy. Modernity defined humans as rational, capable of self-improvement and potential goodness, and established clear boundaries between humans and 'other animals'. From the seventeenth to the twentieth century, animals were treated primarily as a resource for human improvement, so that meat-eating, the use of animals in research, etc. became standard practices. As we move towards postmodernity in the twenty-first century, 'misanthropy' has become a feature of contemporary social life. What he means by this is that humanity is increasingly regarded with a critical eye, seen as a destructive species at whose hand the 'environment' suffers. Franklin contends that animals in late modern society are also associated with a sense of 'risk'. This can be seen in food scares around infected meat, concerns about the preservation of 'wildlife' and the extinction of habitats and, more generally, growing anxieties about environmental pollution that conceptualize humans and animals as subject to a similar threat. Finally, individuals suffer 'ontological insecurity' due to a depletion of family ties, sense of community and neighbourhood with changes in domestic relations (increased divorce rates and re-marriage) and patterns of employment (with 'flexible' labour markets, higher unemployment and less job security). Consequently, they look to relationships with pets to provide stability and a sense of permanence, and a 'feel good' factor (1999: 36) in their lives, because they are pleasing to look at or are subjects of human responsibility. Thus, in late modernity, we are developing 'increasingly empathetic and decentred relationships' with other animal species (1999: 35).

Franklin does not adopt a purely socially constructionist view of animals, however, and there are elements of Benton's critical realism in his position. For example, he argues that it is because pets have some similarities to us (they are animate beings, which respond to external stimuli and often to our company) which enables people to form such strong bonds with them. This said, he still over-emphasizes social changes such as alteration in family structure, and ignores the impact of 'animal rights' on popular perceptions of animals. In addition, while most social constructionists are keen to take account of people's perceptions of events in their research, neither Tester nor Franklin refers to such perceptions, but instead they make assumptions that certain social changes (e.g. in the family) have led to certain practices (e.g. more people keeping pets), and that the reasons people do so is to provide security. As we will see in the section on pets below, this is not necessarily the only explanation for these phenomena.

The politics of animal rights

> Now, comrades, what is the nature of this life of ours? Let us face it: our lives are miserable, laborious and short. We are born, we are given so much food as will keep the breath in our bodies, and those of us who are capable of it are forced to work to the last atom of our strength; and the very instant that our usefulness has come to an end we are slaughtered with hideous cruelty. . . .

> Why do we then continue in our miserable condition? Because nearly the whole of the produce of our labour is stolen from us by human beings. There, comrades, is the answer to all our problems. It is

summed up in a single word – Man. Man is the only real enemy we have.

> (Major the pig, exhorting the animals of Manor Farm to 'rebellion', from George Orwell, *Animal Farm*, 1949: 11–12)

George Orwell might perhaps have been surprised that the sentiments of Major, the old white boar from his satire on Stalinism in the Soviet Union of the 1940s, have been seriously echoed by human agitators for 'animal liberation' over the past thirty years. In one of the earliest 'animal rights' texts, *Animal Rights in Relation to Human Progress*, Henry Salt expected that, to many readers, his 'contentions will appear very ridiculous' (1980 [1892]: xi–xii). The notion of animal rights as a subject of ridicule formed part of the opening chapter of *Animal Liberation*, first published in the mid-1970s by Peter Singer. Singer recalls that one of the early demands for women's rights, Mary Wollstonecraft's *A Vindication of the Rights of Women*, was ridiculed in a satire entitled *A Vindication of the Rights of Brutes*. This contended that the extension of rights claims to women was flawed, as the same logic could extend rights to non-human animals, and this was patently absurd (Singer 1990: 1). Singer's book used many of Salt's arguments, but whereas Salt's treatise fell into immediate obscurity, Singer's became both a bestseller and, in some cases, a handbook for political activists concerned with animal welfare.

Theories of animal rights

Tester contends that there are two types of argument deployed for animal rights, based either on respect for difference or on similarity (1991: 94–169). Those emphasizing 'difference' between humans and animals have argued that humanitarianism was demonstrated through 'right treatment' of other animal species. Those emphasizing similarities between humans and animals have a romantic view of 'true nature' which involves closer relationships with animals (1991: 145). The work of both Singer and Salt combines what Tester sees as rationalist and 'romantic' discourses. Singer argues that we both express human morality (thus improve ourselves) and improve the situation of animals in the acceptance and application of animal rights. He argues that the attribution of 'rights' to animals is necessary because, in liberal philosophy, those who have 'rights' are free do as they please, as long as they do not impinge on the freedom of other rights-bearing individuals. If humans are not to mistreat animals and use them for their own ends, then animal rights will be necessary to impose ethical restrictions on human conduct.

Singer was the first to use the political language of 'liberation', 'oppression' and 'discrimination' with reference to human relations with animals. Animals are oppressed, exploited and discriminated against because they lack rights in a society that is 'speciesist'. 'Speciesism' (a corollary to racism and sexism) is the belief that humans are entitled to treat members of other species in ways in which it would be deemed morally wrong to treat other humans (Ryder in Rollin 1981: 89–90). In short, speciesism is discrimination based upon species membership and, as Cooper (1995: 138) points out, Singer *et al.* see this as a form of outmoded 'unreason'.

Singer (1990) and Tom Regan (1988) base part of their case on material drawn from the study of animal biology, which indicates conscious awareness in non-human animals. They contend that animals have rights because they are 'sentient', that is, they are capable of experiencing pain, suffering and pleasure. Mary Midgley agrees with Singer (1981, 1985) that we should extend the principle of moral worth from humans to other animals, and treat 'all sentient beings as inside the moral community' (Midgley 1983: 89). Singer, Regan, Midgley and others tend to use an image of moral progress and improvement, thus creating, as Richard Ryder puts it, 'an ever-widening moral circle' into which new 'classes of sufferers . . . are drawn' (Ryder 1992: 3). However, expanding the 'circle' of ethical concern from different groups of suffering humans to 'animals' can be problematic, as it ignores the vast differences across the multivariate animal species (Cooper 1995: 141). Singer agues that all vertebrate animal groups (i.e. mammals, birds, fish, reptiles, amphibians) are 'sentient' and can have rights attributed to them. Regan has a more limited conception of which animals deserve rights, contending that to have rights, animals must be the 'subjects of a life' (1988: 367), and that this only applies to animals (specifically mammals) with certain similarities to humans.

While Singer and Regan stress human similarities to other higher mammals, Midgley (1983) contends that arguing for rights for animals is different from arguing for the extension of rights claims from one group of humans, for example, white, middle-class men, to women of various social locations of class and race. This is because animals 'are not just animals. They are elephants or amoebae, locusts or fish or deer' (1983: 19). Midgely thinks species barriers are permeable and less certain than they are currently perceived to be, however, and dismisses the arguments of 'animal rights' detractors who contend that animals are not entitled to rights because they do not have interests. Raymond Frey, for example, argues that animals do not have interests because they do not think or speak. Consequently, we cannot conceive of them as individuals who might bear rights (Frey 1983: 109). As Midgley points out, on this line of argument, we would also preclude rights to humans if they were babies, dumb and perhaps deaf or had some form of disability that affected the ability to communicate through speech (1983: 56–60). Frey fails here to account for the similarities some animals may have with some humans and exaggerates the human–animal divide.

Singer exaggerates the similarities, and continually makes comparisons between 'speciesism' and intra-human forms of discrimination:

> Racists violate the principle of equality by giving greater weight to the interests of their own race. . . . Sexists violate the principle of equality by favoring the interests of their own sex. Similarly, speciesists allow the interests of their own species to override the greater interests of members of other species. *The pattern is identical in each case.*
>
> (1990: 9, my emphasis)

He goes on to say that 'Most human beings are speciesist', and that through everyday practices such as meat-eating, most of us are complicit in animal abuse. He argues that the 'like suffering' of animals is of equal value, and that, despite species difference, we can account for differential treatment while also endorsing the principle of equality. The extent of species difference, however, means that in some cases the argument about sentience and suffering does not apply, at least not in the same way. Midgley (1983: 104) argues that we should not play down the differences between animals as we should the differences between human beings in opposing sexism and racism. To imply discrimination against all animals by all humans does not account for intensity of cruelty inflicted, or species difference. Singer also ignores the complexity of formations of gender relations and ethnic structuring of societies, and the ways in which these forms of social inclusion and exclusion vary within certain societies and alter over time.

This said, Midgley underplays some of the commonalities in the treatment of certain humans and certain animals at particular historical junctures. For example, Marjorie Spiegal (1988) makes a careful comparison of the slavery of animals and that of black humans. Eco-feminists have examined similarities and differences in the treatment of animals and women for example, in domestic violence situations (Adams 1995), and the application of new reproductive technologies (Corea 1985). The continued embeddedness of animals in human communities is the basis of

Benton's (1993) critique of animal rights politics, and Franklin (1999: 175) also argues that separation of human from animal cultures in order to avoid animal abuse is 'out of kilter with the general direction of change in recent years'.

A further strand of criticism of Singer and Regan is their entrenched rationalism (Singer 1990: iii; Regan 1988: 94). Critics argue that, in developing 'interspecies justice', we need to practise 'sympathetic identification' with animals (Johnson 1995: 166). Josephine Donovan claims that 'womanish' sentiment is being criticized in trivializing an emotional response to animal abuse (1990: 351). Together with Carol Adams (1995) she has developed an approach to animal rights rooted in the social context of women's caring traditions. This is based on the notion of concern for sentient creatures rather than identifying 'rights' animals might have. Unfortunately, Donovan and Adams do not account for different kinds and levels of treatment and care, concurring with Suzanne Kappeler that approaches establishing hierarchies and boundaries of difference between species are 'discourses of domination' (1995: 331). Diane Antonio (1995) provides a compromise by suggesting that an approach to human–animal relations that is rooted in a notion of 'care' should be supplemented by a respect for diversity and difference occurring in the non-human natural world. So, we need to understand diversity and differentiation by species, and when we 'care' for animals, must 'respect' diversity and tailor our notions of appropriate human treatment to the situation of differing species.

How might an ethic of care or respect, or an appreciation of 'rights', manifest itself in policy terms? While it may not be the most sophisticated analysis, elements of Singer's recommendations have been adopted by many animal rights organizations. These demands are that people should stop rearing animals for meat and adopt **vegetarianism** or **veganism**, and that hunting, vivisection, and the trade in animal goods such as fur should be abolished. We will briefly consider how different social movement organizations have articulated such demands below. Singer's enthusiasm for extending rights to animals is undiminished, despite the problems suggested with this individualist, rights-based approach (Benton 1993). His latest involvement is with the 'Great Ape Project' which aims to obtain a United Nations 'Declaration on the Rights of Great Apes', which would furnish them with a right to life, to take part only in benign experiments, and not to suffer 'cruel or degrading treatment' (see Holden 2000: 54).

Animals and social movement organizations

Tester characterizes animal rights groups as cults with fundamentalist convictions and purificatory (and puritanical) rituals such as vegetarianism/veganism, and abstinence from contact with animals such as 'pet' keeping (1991: 178). However, as Franklin (1999) points out, Tester's assumptions about animal rights activists are based on very limited empirical evidence, and a few press releases from certain 'extreme' animal rights groups is hardly representative research. Being interested in animal rights as a religious cult rather than a social movement, Tester ignores the range of organizations with different objectives, tactics and membership compositions. Unlike Franklin (1999), I do not think it is helpful to consider only those who adopt a more radical 'animal rights platform'. Social movement organizations change in policy and orientation over time, and have a collective impact on popular opinion about appropriate relations between humans and animals. This brief survey concentrates on some British groups, with some reference to the international organization of animal rights protest.

Britain has a long history of animal welfare campaigning, beginning with nineteenth-century organizations such as the Society for the Prevention of Cruelty to Animals, which was formed in 1824 (becoming the Royal Society in 1840, Franklin 1999: 184), a constitutional pressure group focusing on parliamentary lobbying. The RSPCA has been influenced by animal rights debates and, in the 1980s for example, bowed to internal pressure from its membership to adopt an anti-hunting stance (Garner 1993a). Other organizations have historically articulated more radical demands, such as the British Union for the Abolition of Vivisection (BUAV, founded 1898), which calls for the abolition of all animal-based medical, military and other scientific research, and has now diversified its aims and objectives as part of the Animal Alliance.

According to Byrne, there are three strands to animal rights protest in Britain: opposition to blood sports, particularly the hunting of mammals; opposition to farming practices particularly intensive production and live exports; and opposition to vivisection (1997: 184–5). Each of these strands contains groups advocating direct action and prepared to break the law, and those engaged in constitutional political action. Thus, for example, he cites the League Against Cruel Sports (founded 1924) as a conventional pressure group opposed to hunting, and the Hunt Saboteurs Association (founded 1964) as the direct-action wing of the anti-hunting lobby. However, while this schema may impose some categories on the plethora of groups, it does not necessarily give an accurate guide to their politics. For example, Byrne 'classifies' the Animal Liberation Front (ALF, founded 1976) as an anti-vivisection organization. This ignores its formation by a group of hunt saboteurs who began to damage the vehicles of those participating in hunts. The ALF has 'liberated' animals from fur and factory farms and undertaken arson attacks on vivisection laboratories and retail outlets such as McDonald's. Such damage to property has become a subject of anti-terrorist police action as a result of successful lobbying by sections of the scientific community.

Many groups are concerned with a variety of animal welfare and protection issues. Compassion in World Farming (CIWF, founded 1967) is concerned with welfare issues in meat, milk and egg production, and seeks to end the worst extremes of factory-farming practice and the trade in live animal exports. CIWF was a significant organizing force behind the series of demonstrations against live exports in 1995 at British ports and airports that attracted considerable media attention. Animal Aid (founded 1977) has a radical programme of non-violent direct action, based directly on Singer's *Animal Liberation*. It has launched publicity campaigns against factory farming, engaged in undercover research into the farming, transport and slaughter of meat animals, and into vivisection practices, and advocates vegetarianism/veganism, a ban on animal circuses and zoos, and a radical reconceptualization of our current attitudes towards animals.

International organizations seem to be increasingly significant, and are interested in a range of issues. The World Society for the Protection of Animals (WSPA, founded 1981 through the merging of two previously existing groups) is concerned with habitat depletion and loss of species diversity, and the exploitation of animals in zoos, etc. and in situations of political and/or ecological crisis such as warfare. With 340 societies in 70 countries and consultative status with the United Nations (Franklin 1999: 185), the WSPA can be seen as an influential political player. More radically, PETA (People for the Ethical Treatment of Animals, founded in 1980) has over half a million members, concentrated in Europe and America. PETA contends that animals are not 'ours' to eat, wear, experiment on, etc., and it engages in high profile media campaigns to attempt to popularize animal rights issues particularly among young people. It engages in non-violent direct action on issues such as factory farming and animal experimentation (involving demonstrations and 'rescues'), lobbying (where it has been particularly successful in the United States), public education strategies through schools and colleges, and advertising and celebrity media events.

At the time of writing, there is heated debate in Britain over the Labour government's proposals to introduce legislation to ban the hunting of mammals such as deer and foxes with hounds. Since the 1997 General Election, the British government has clearly held the view that, in attempting to ban hunting, they are reflecting the majority of British public opinion. Animal rights and welfare activists see fox hunting as a symbol of the political acceptance of cruelty to animals, and have been effective in lobbying MPs who have recently voted decisively for an outright ban (*The Guardian*, 19.3.02). Whether the government will invoke its constitutional powers to overrule the pro-hunting upper chamber is a moot point, but it is apparent that the political conception of animal rights, or at least animal welfare, now has some level of mainstream political support. Comparative research on attitudes to pets and other animals in the United States and Germany indicates a high level of support for avoidance of cruelty towards animals and a strong identification with domestic 'pets' (Serpell 1994: 173). Thus animal rights has become a salient political issue, reflecting the changing relations between humans and animals suggested earlier in this chapter.

Animals as food production units

Until recently there has been little attempt to understand, sociologically, the relation between animals and human culture in terms of food products. Even recent literature on food consumption in the West (Lupton 1996; Warde 1997) has ignored the possibility that the widespread consumption of meat, and the processes of livestock rearing to produce muscle, skin, milk and eggs for human consumption tells us something about human culture's relationships with non-human animals.

One of the initial attempts to analyse human–animal relations sociologically in respect to food production and consumption was Jeremy Rifkin's *Beyond Beef*. Rifkin contends that beef is key to a social structure observable at particular historical periods in certain societies: the 'cattle complex' (1994: 3). Rifkin argues our notions of 'civilization' in Western societies have been fundamentally shaped by our relationship to animals, particularly in the cases of the United States and much of Europe, to cattle. He examines the history of the human–bovine relationship and contends that changes in social relations involve human relations with animals. For example, he sees the rise of European Christianity as instrumental in transforming cattle from objects of pagan worship to objects of contempt, and he discusses the significance of cattle in the establishment of market economies in Europe, with cattle being the oldest form of mobile wealth. In more recent history, Rifkin notes that meat production and consumption was linked to forms of social hierarchy around class, race and gender. In the United States, for example, white colonialism resulted in dramatic change for both the indigenous peoples and indigenous animals, with extermination of both bison and Amerindian people, and their replacement with white homesteaders who bred domestic cattle. He also notes that levels of meat consumption and the types of meat consumed, have been – and continue to be – linked to class position and gender (1994: 241).

Franklin (1999: 40) describes the development of intensive animal-rearing in the last quarter of the nineteenth century in North and South America and Australia, arguing that the rationalization of factory-style meat, milk and egg production increasingly meant that livestock-rearing, marketing and slaughter was confined to rural areas. Western affluence and cheap intensive production meant that meat-eating increased and was a practice recommended by governments. So, human contact with animals, in the form of their dead flesh, increased as the twentieth century progressed, and, until the 1970s, there was the continued cultural popularity of meat food, due to its association with high social status and wealth. However, alongside such increased consumption came an increased distaste for the process of slaughter and butchery, which has led to the popular presentation of meat as hardly recognizable as an animal carcass.

Franklin contends that, throughout the twentieth century, two very different practices could be observed: the increased exploitation of animals as meat, milk and egg production units through intensification of farming (largely obscured from public view), and the romanticism of the farmyard as a rural idyll, in children's books for example (1999: 127). Animal welfare groups have attempted to raise public awareness of the material production of meat by campaign images that show the conditions of factory farming and the process of **disassembly**. Such representations question romantic images and perceptions of farming, but are not part of the popular representation of either animals or of meat (see Figures 6.1 to 6.4).

In their research into farming practice across developed and 'underdeveloped' countries, Alan Durning and Holly Brough (1995) argue that the move to 'factory-style livestock industries' in developed countries in the last fifty years has resulted in widespread environmental

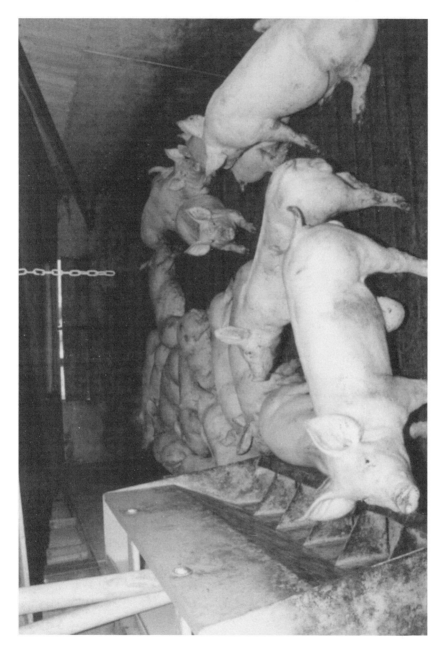

Figure 6.1 Intensive production I: pigs in a fattening pen
(*Source:* Compassion in World Farming)

Figure 6.2 *Intensive production II: battery hens*
(*Source*: Compassion in World Farming)

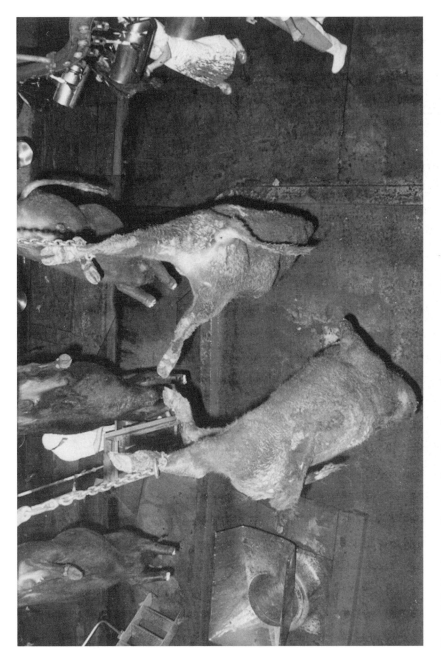

Figure 6.3 *Rarely seen images of meat production I: cattle conveyor belt*
(*Source*: Vicky Aldaheff)

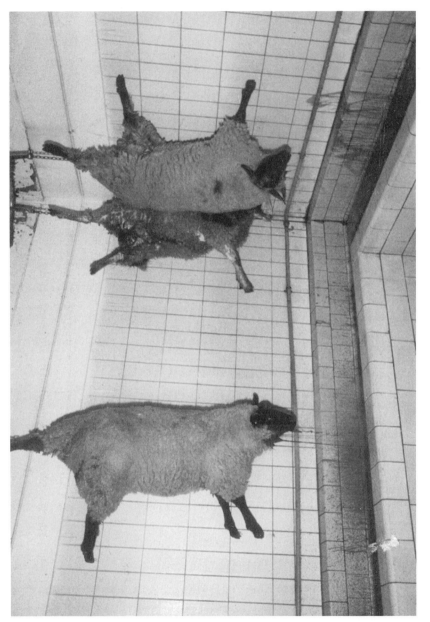

Figure 6.4 *Rarely seen images of meat production II: sheep on the slaughter line*
(*Source*: Compassion in World Farming)

side effects, along with the 'diseases of affluence' assisted by meat-rich diets (1995: 149). The subsidization of Western practices in developing countries has put the ecologically effective livestock economy of some regions out of alignment with social need and desirable environmental practice, resulting in deforestation, contributing to global warming, desertification and excessive water usage (1995: 153–9).

Meat-eating differs quite markedly cross-culturally. For example, in Japan much fish and shellfish is consumed, and red meat forms a very limited part of the diet. In the key beef-producing nations, such as the United States and Argentina, beef is an important food, and meat remains an assumed centrepiece of every meal. In countries that have high levels of lamb production, such as New Zealand and Australia, lamb consumption exceeds that of other meats, and lamb, mutton and goat are important foods in the arid regions of the Middle East and southern Mediterranean. Due to cultural taboos on eating cattle, in the North Indian 'Hindi belt', plenty of dairy products, but a minimal amount of beef, are consumed (Franklin 1999: 146–7). Despite these diverse levels of consumption, there are general patterns of change in meat-eating, particularly in the Western world.

Most sociologists writing about food consumption concur that we have witnessed a growth in vegetarianism, usually viewing vegetarianism as an eccentric behaviour related to postmodern social anxieties, especially risk surrounding certain animal foods (Lupton 1996). There is a trend towards consuming more fat-rich and processed foods, which Rifkin (1994) refers to as 'deconstructed meat' – the US hamburger being his prime example. Franklin (1999: 148) thinks we are witnessing the break-up of popular modern meat culture, with the increase in vegetarianism and the perception among consumers that meat consumption is among the possible risks of late modern society. He argues that meat-eating has not suffered a decline *per se*, but that there have been differential fluctuations. While we witnessed an overall decline in consumption in the 1980s, in the 1990s there was a slight increase. There has been a change in the form of consumption, however, with consumers now opting for meat(s) they consider to be low risk, such as 'game'. In the developed world, BSE (bovine spongiform encephalopathy) and its human incarnation nvCJD (new variant Creutzfeldt-Jakob disease) and salmonella have been associated with intensive production. Class division is also apparent in the West, and the class division of meat consumption has shifted. Higher class groups have reduced meat consumption levels significantly, and poorer social groups, who have historically been relatively excluded from high meat consumption due to the cost, are consuming more due to cheap intensively produced meat.

Some sociologists have examined rituals surrounding contemporary meat consumption, often drawing on structural anthropology. Of particular interest has been the connection between meat-eating and the social construction of masculinity (Twigg 1983; Bourdieu 1984; Charles and Kerr 1988). Pierre Bourdieu (1984: 190–2) has asserted that in French popular culture there is the belief that fish, fruit and vegetables are light feminine foods that will not prove sufficiently 'filling' for men, who require the supposed energy-giving properties of red meat. Nicky Charles and Marion Kerr's (1988) study of meat-eating in British families revealed a strong belief among both men and women that men should consume the most meat, and Julia Twigg (1983: 23) suggests that the cultural symbolism of meat is associated with masculine strength and sexual potency. Carol Adams (1990) makes some interesting observations about contemporary meat-eating culture. She uses the term 'absent referent' (1990: 40) to describe how we usually refer to meat with a word which obscures its animal origin. To say 'pork' rather than pig-meat (1990: 26) distances us from the animals we eat. Adams argues that meat is assumed to be 'male food', and is represented in popular culture in a way that both sexualizes and feminizes it. The cultural presumption of meat as masculine food may explain the higher incidence of female vegetarians, which is so marked among young British women that, in the 1990s, the Meat and Livestock Commission concentrated on wooing back lapsed female consumers through advertising campaigns (Cudworth 1998).

For Franklin, animal rights arguments do not dominate changing patterns of meat consumption. I do not agree however that discourses on animal rights also 'do not dominate contemporary vegetarian attitudes' (1999: 162). Among existing vegetarians, at least in the UK, animal welfare is the prime reason given for abstinence from meat-eating (interview in Cudworth 1998). Some sociologists over-sociologize the impact of risk associated with meat-eating, not taking on board Beck's (1992) point that environmental risks are both perceived and real. Deborah Lupton (1996), for example, cites concern with BSE as a perceived risk only, arguing that consumers are responding to media presentation of beef as a risk. Alan Warde's (1997: 94) analysis of food, consumption and taste argues that complex and divergent factors affect our choice of food. Although there has been a shift to 'healthy' (often vegetable) food, this is contradicted and countered by a culture of 'extravagant' and indulgent food (which is likely, in the British diet, to be meat based).

Modernity has altered our relationship with 'food animals' by intensifying production and separating most people from the meat production process. Modern societies, until very recently, were associated with relatively high levels of meat consumption and a cultural ethos (differentiated by social hierarchies such as gender and class) which prescribed it. In late modern societies, there is a critical reflexivity about meat and changing patterns of consumption, albeit that these are differentiated cross-culturally and across social groups. There is also anxiety around modern food production methods as a result primarily of conceptions of health 'risk' in relation to meat, but perhaps, to a degree, because of the influence of animal welfare campaigning as well.

Animals as household companions

Thomas (1983) suggests that there are a number of criteria by which we can define 'pets' or, as animal rights activists prefer, 'companion animals'. 'Pets' are distinct from other domestic and from wild animals because they share our homes, we give them names, and we (at least for the most part) do not eat them. In his historical account, Thomas describes pet-keeping as very much an affair of the upper and expanding middle classes, who fed and often treated companion animals 'better than the servants' (1983: 117). By the late eighteenth century however, Thomas contends that the practice of pet-keeping (particularly of dogs) had filtered down to all levels of society (1983: 105). Harriet Ritvo has argued that the range of animals kept as domestic companions in Western countries has become more diverse as the twentieth century has progressed (1987: 177). However, from the nineteenth century, she argues, pet ownership has been divided by class in terms of the species of animals kept. Horses, exotic pets and pedigree dogs and cats were a middle- and upper-class preserve, whereas cross-breed cats and dogs, rabbits, mice and caged birds were associated with the working classes (1987: 82).

From the 1960s, pet-keeping has increased significantly across the countries of Europe, America and Australia, and has also grown in countries less renowned for pet-keeping, such as Japan (Franklin 1999: 89–90). Sales of pet food, pet health and pet care products has risen, and the range of such products expanded. The nature of pet-keeping itself has also changed, in that the popularity of 'rare' or exotic animals such as reptiles has spread across classes, while certain practices which may be seen as morally dubious from an animal welfare perspective, such as the keeping of caged birds, has declined, as has the popularity of certain breeds, such as 'toy' dogs. Pets have become valued as friends and companions rather than playthings. Even the phenomenon of 'cyberpets', as a 'surrogate' for a desired activity, may indicate the popularity of 'real' pet-keeping. Recent evidence of increasingly close relations between animals and human societies includes the establishment of such institutions as pet cemeteries, funeral rituals and psychological treatment, giving pets human names and gourmet pet food, which seems to suggest that pets are becoming 'quasi family' (1999: 94–5).

While most of the few commentators agree that changes have taken place, the reason for such changes, and the understanding of pet–human relations differs markedly. Strong constructionists such as Tester (1991) contend that pet–human relations are projections of idealized relations between people, so viewing pets as integrated with human households is a preposterous anthropomorphizing of animals. Franklin (1999) sees the keeping of certain pets, particularly dogs, as related to the construction of individual self-identity; they are partly a 'lifestyle choice':

> in Britain and Australia for example, pit bull terriers with brass-studded leather collars have become familiar accessories to the physically tough 'bloke', apricot toy poodles have been popular among effeminate gay men. Standard poodles, Afghans, Pomeranians and salukis are feminine dogs, associated particularly with the urban, wealthy and young. Similarly, pugs, schnauzers and chihuahuas are associated with older middle-class women.
>
> (Franklin 1999: 100)

Franklin also wants to account for 'real' changes in people's relations with pets. He suggests that pet-keeping, like other aspects of human–animal relations, has become 'decentred' (1999: 86). It is not simply that animals are more anthropomorphized. Pet keepers are decentring – that is, learning to think about and trying to understand the needs of others, and increasingly thinking about animals as animals (not human substitutes), and looking for the similarities and differences between themselves and various species of pet animal. Jossica Newby (1997) further argues that the shared communications between people and their 'pets' means that pets *are* family.

Franklin (1999: 98) argues that sociological texts on the 'family' should take account of relationships with pets. While I agree this is often an omission, there is some work undertaken by eco-feminists that has made connections between what some sociologists call the 'dark side' of family life – domestic violence and violence towards domestic pets. Adams's (1994, 1995) more recent work looks at how threats to neglect, harm or kill family pets are used as threats against women and children. This makes grim reading, and estimates by animal welfare charities in Britain, indicate that the number of companion animals abandoned and mistreated seems on the increase. However, I think Franklin is right to suggest that a more reflexive and animal-centred conception of pet-keeping is generally establishing itself in Western societies.

Animal entertainment

In Western cultures, animals are very much part of popular culture. In entertainment such as television and film, and in advertising, animal imagery abounds. More limited is the human–animal interface in leisure pursuits involving animals as 'game' and humans as hunters. Angling remains a widespread and popular pursuit, and the hunting of wild mammals and shooting of birds persists, despite its appeal to a more limited range of social groups. In contrast to the desire to look at animals, Franklin (1999: 105) sees hunting and angling as 'enigmas in modernity', by which he suggests these practices buck the trend of human–animal relations. A more sympathetic treatment of animals has been characteristic of social development over the last century with an increasing amount of 'anti-cruelty' legislation and expansion of pet-keeping. Hunting and angling can be seen as 'anti-modern' in the sense that they are 'countryside pursuits' involving a pre-modern relationship to nature, combining elements of conservationism with knowledge of natural processes such as seasons, tides and wild animal behaviour.

Hunting cultures

Franklin (1999) makes an interesting comparison of hunting cultures and practices in Britain and the United States, and contends that the social composition of animal-based sports in these two countries is very different. In the United States, hunting is part of national identity and is associated with the social construction of masculinity. Hunting, and especially angling, is almost exclusively a white pastime, and both are divided by class with reference to the kind of animal sought (1999: 109). A new development in the commercialization of hunting is the 'game ranch', where hunters may pre-order exotic animals they wish to hunt as part of the ranch stock, and pay for the animals they kill. In Britain, both hunting and fishing rights were traditionally the preserve of the landowning classes. While during the twentieth century fishing has become a more widespread pastime across the class structure, hunting, particularly of deer and foxes, has remained overwhelmingly an elite activity, and may soon no longer be legal.

In explaining the persistence of hunting as a leisure activity, Franklin (1999: 117–23) suggests four features of hunting culture: anti-modernism (a romantic notion of being in harmony with nature), neo-Darwinism (hunting is a natural attribute of our species), conservationism, and the construction of masculinity. While he notes that hunting and fishing are almost exclusively male activities, with a culture of male fraternity based on a clear separation of men from domesticity, he sees this as an 'under-researched area'. However, he ignores the eco-feminist literature that elaborates upon the importance of hunting as an element of masculinity. Andrée Collard (1988: 33) sees the myth of 'man the hunter' as part of the discourse of contemporary masculinity in modern societies, and suggests that the romanticization of nature is often incorporated into hunting cultures. Collard (1988: 52) does not see hunting as associated with 'men' *per se*, but with certain kinds of men adopting a traditional form of masculinity (see also Comninou 1995). In Britain, hunting culture is associated with men, but also structured by class (shooting for example, being less of an elite preserve than fox or deer hunting). Certainly hunting culture has a particular view of animals: as quarry for human sport and as 'pests'. The hunting aesthetic is pre-modern in that, contrary to the cultural representation of modern meat, carcasses are celebrated and often photographed as trophies (see Figure 6.5).

Franklin is right to suggest that hunting contradicts the generally more benign attitudes towards animals in modern societies. It is interesting to note however, that in Britain, where hunting in some of its forms is under threat of legal restriction, the pro-hunting lobby has increasingly presented itself as conservationist. This shift in emphasis, and indeed the political motivation to restrict hunting, can be seen as evidence of the kinds of changes in human–animal relations discussed above.

The 'zoological gaze'

Nature documentaries have increased significantly in the airtime they occupy, and have diversified in what is arguably an 'eco-friendly'

Figure 6.5 *The hunter's-eye view*
(*Source*: John Monger)

direction. For example, programmes about plants and micro-organisms have been a feature since the late 1970s. The popularity of animals in cartoons and children's films continues. Those made by the Disney Corporation have historically tended to portray animals as innocent or good and humans as unpredictable, careless or cruel. In Britain, programmes about 'pets', vets and even pet therapists, crowd the viewing schedules. Images of 'aesthetic' animals pervade popular culture and are increasingly used to market or simply to decorate products such as cheques and credit cards. Franklin (1999: 62) refers to the practice of looking at animals, and the desirability of looking at animals as the '**zoological gaze**', and it does seem that gazing at animals is both increasing popular, and also different in content in recent times.

A key site of the zoological gaze is, obviously, the zoological garden, which has not necessarily declined in popularity, but has certainly altered in form. The earliest zoos, in London and Paris, were spin-offs from royal animal collections or 'menageries'. Alongside this, most working-class people in the West would have seen exotic animals through travelling menageries or circuses (Thomas 1983). The zoological gaze developed around the notion of the exotic and the grotesque, with wild animals seen as dangerous and strange 'others', requiring prison-like accommodation. Thus it may not be coincidence that London Zoo developed from the royal menagerie housed in the Tower of London (Bostock 1993). The humanitarianism of the Enlightenment was also applied to non-humans; for example, French revolutionaries 'freed' the animals in the royal menagerie at Versailles and, post-revolution, zoo functions included education as well as presenting a 'spectacle' for the viewing public.

The idea of zoo animals as entertainment for (mostly young) humans was to emerge as the predominant view for much of the twentieth century. Animals were seen as 'actors' for human entertainment rather than objects for us to observe – so camels gave children rides and primates had (and in some cases continue to have) 'tea parties' (Bostock 1993: 34). As the emphasis on entertainment increased, so animal enclosures changed to favour the keeping of animals (even from warmer climates) outside in large enclosures separated from the public often by moats rather than bars. Animals were cast as actors for human entertainment, rather than objects for human observation.

The animal rights movement emerging in the late 1970s changed the perception of zoos, and Singer (1990), for example, defined zoos as prisons. Franklin (1999: 73) sees this critique of zoos as extreme and argues that it marginalizes the politics of animal rights. While abolishing zoos may not be in line with public opinion in terms of the desire to watch animals, the critique of zoo practices did strike a chord with the increasingly sympathetic attitudes to animals. Franklin argues that zoos have become 'decentred' (1999: 74) since the 1970s, and sees this as a result of the shift in human–animal relations due to ontological insecurity. However, elements of the animal rights agenda have been taken on board by zoos as a means to preserve their public sympathy – zoos represent themselves as scientific research establishments concerned with the protection of rare and endangered species. 'Good practice' now involves zoos entertaining their animals rather than the human visitors, with television, games, hiding food to encourage foraging and recreating 'natural' habitats, ensuring privacy from spectators and establishing close relations with keepers.

In Western societies, the use of animals as a form of 'entertainment' has increased in popularity, but changed in form. While some arguably cruel practices such as bull fighting, shooting and hunting continue as leisure pursuits, others have been prevented by law (such as bear and badger baiting). If we take zoos as a key example of 'animals as entertainment' we have seen a radical alteration in the principles behind these institutions in the past century.

Conclusion

Humans have always had relations with other animal species, and, in modern societies, we continually interact with animals in our everyday lives. In the last century, dominant conceptions of animals and common practices towards animals have altered significantly, and in the areas of human–animal interaction examined here, entertainment, food production and pet-keeping, change is evident. There has been increased concern about the welfare of food animals, and about the implications for human health of intensive farming methods. Food consumption practices have altered, with an increase in vegetarianism, and increased consumption of both 'free-range' and processed meat products. More people are keeping different kinds of animals as 'pets', but, as with the viewing of animals as popular entertainment, attitudes to animals have become more 'decentred'. This means that the needs of different animal species have become an important criterion, whether those animals are kept as domestic companions or in zoos. Even when we consider the use of animals in hunting and angling, we can see not only different conceptions of human–animal relations, but also the increased influence of arguably more benign attitudes with the dominance of conservationism.

Sociologists have different explanations for the nature and extent of these shifts in human–animal relations. Some feel that the nature of contemporary society has led to anxieties about the implications of modernization, so that we are becoming more critical and reflexive about a range of relationships, those with animals being one arena of consideration. Others argue that real environmental changes, such as habitat depletion and climate change have forced humans to reconceptualize our relations with animals, and that we are developing less human-centred attitudes as a result. Certainly, the arguments put forward by the animal rights movement have had an impact in altering certain aspects of human–animal relations in modern society and can be seen as part of this 'decentring' process.

Key points

● The concept 'animal' is a social construct and an examination of the history of human–animal relations in Western Europe, even in the relatively short space of the last two centuries, indicates that human relations with animals, and our conceptions of animals, change significantly over time.

● Sociologists have different ways of theorizing human relations with animals. Strong social constructionists see human–animal relations as reflections of relations between human beings, realists see such relations as both socially constituted and shaped by 'real' environmental factors.

● 'Animal rights' has been one way of conceptualizing appropriate relations between humans and other animals, and as both a set philosophical arguments and a political stance, animal rights has increased in influence in modern societies.

● While most people eat meat and other animal products, and wear leather and wool, there are increasing levels of anxiety around the production and consumption of animal products, with some changes in dietary habits.

● In modern societies, the number of people keeping domestic companion animals has risen, and the range of species kept as 'pets' has diversified. Sociologists have different explanations of why these developments have taken place. Some argue that pet-keeping is increasing due to rising levels of insecurity due to social change, others see this as a result of more positive understandings of different animal species.

● The use of animals in sports such as hunting and fishing remains popular among a significant minority of the population. Different kinds of hunting and angling have followings from different social groups. Both hunters and anglers have tended to stress conservation as a justification for their sport in the late twentieth century.

● Many people like to watch animals on television or 'live' such as in zoos. However, the way we watch animals and the institutions of zoos have changed to become less 'human-centred'.

Further reading

Useful for a number of areas, but particularly historical change in human–animal relations and the historical development of animal-rights, is Keith Tester's *Animals and Society: The Humanity of Animal Rights* (London: Routledge, 1991). Perhaps the key historical text is Keith Thomas's *Man and the Natural World: Changing Attitudes in England 1500–1800* (London: Allen Lane, 1983). Also useful is Harriet Ritvo's *The Animal Estate: The English and Other*

Creatures in the Victorian Age (Cambridge, MA: Harvard University Press, 1987), or there is a shorter summary of some of her arguments in her chapter 'Animals in Nineteenth-Century Britain', in A. Manning and J. Serpell (eds) *Animals and Human Society* (London: Routledge, 1994).

As a general overview of a wide range of the literature I would recommend Adrian Franklin's *Animals and Modern Cultures* (London: Sage, 1999); the third chapter in particular is useful for theoretical explanations of human–animal relations. Good examples of the differing sociological perspectives of postmodernism and realism respectively, are Tester (1991, above) and Ted Benton's *Natural Relations: Ecology, Animal Rights and Social Justice* (London: Verso, 1993).

On the philosophy of animal rights, Peter Singer's *Animal Liberation* (New York: Avon Books, 2nd edn 1990) is a lively read, and important due to its impact on political activists and academic debate. Alternatively, there is Tom Regan's *The Case for Animal Rights* (London: Routledge, 1988). Mary Midgely provides some interesting arguments in her concise and readable *Animals and Why they Matter: A Journey Around the Species Barrier* (Harmondsworth: Penguin, 1983). For a perspective which is clearly opposed to animal rights arguments, see R.G. Frey, *Interests and Rights: The Case Against Animals* (Oxford: Clarendon, 1980).

Franklin (1999, above) provides an overview of the literature on various sites of animal–human relations, and there are various relevant chapters in the collection *Animals and Human Society*, edited by Manning and Serpell (London: Routledge, 1994). For a more detailed account of zoos, see Stephen Bostock, *Zoos and Animal Rights* (London: Routledge, 1993); on pets, see John Serpell's *In the Company of Animals* (Oxford: Blackwell, 1986) and his chapter in Manning and Serpell (1994, above). For an impassioned and original history of human relations with 'food animals' see Jeremy Rifkin's *Beyond Beef: The Rise and Fall of Cattle Culture* (London: Thorsons, 1994), and for contemporary material on factory farming, Chapter 3 of Singer's *Animal Liberation* (above). Julia Twigg's chapter 'Vegetarianism and the Meanings of Meat', in A. Murcott (ed.) *The Sociology of Food and Eating* (Aldershot: Gower, 1983) is an interesting read in a generally useful collection. An important feminist approach to meat and popular culture is Carol Adams's *The Sexual Politics of Meat* (London: Polity, 1990).

There are many websites devoted to animal welfare and animal rights, including these British examples:

Animal Alliance: www.animal-alliance.org/indexhtml

Compassion in World Farming: www.ciwf.org.uk

League Against Cruel Sports: www.Lightman.co.uk/lacs/

World Society for the Protection of Animals: www.whisper.org.uk

World Wide Fund for Nature: www.wwf.org.uk

Discussion questions

1 What were your last three interactions with an animal? What can these interactions tell us about contemporary forms of human–animal relations?

2 Should the United Nations attempt to guarantee certain 'rights' for primates?

3 Are human–animal relations changing as most sociologists of this subject suggest? Most argue these changes mean contemporary societies are becoming less human-centered. Is this desirable? What might be the practical consequences of such changes in the next 20 years?

Conclusion: environment and society

This book has introduced concepts, theories and issues raised by the sociology of the environment. It has taken a broad sweep in suggesting that although sociology has been nature-phobic until late, it has much to offer. I have included concepts and theorists from this nature-phobic legacy because I think many established sociological ideas can be modified, shifted and usefully applied in trying to better understand relationships between environments and societies.

I have also suggested that while sociologists can tinker with their established concepts, theories and research practices, they might also do well to open themselves up to those from other disciplines, and consider more interdisciplinary ventures in empirical research and theoretical development than, for the most part, they have thus far.

This is not an uncontroversial view. As we saw from Chapter 1, there have been heated debates between those using what I see as a more traditional sociological position which sees nature as a series of contested social constructs, and those advocating a potentially transdisciplinary realist position. In the latter, the environment is conceived as multifarious complex organisms and entities with their own properties and powers, which interact in diverse ways with social processes and develop and change in interaction with them. Co-constructionists have combined elements of these approaches in arguing that humans interact with objects and organisms in complex networks, characterized by co-constructionism. When the different varieties of these positions are considered, they may not be so diametrically opposed as it may at first appear. While my own preference is apparent, I hope to have demonstrated that all these approaches have something to offer those seeking to better understand relations between societies and their environmental contexts.

Chapter 2 outlined some varieties of environmentalism, with the focus on how 'green' thinkers understand the links between prevalent beliefs and values, social institutions and differences, and the 'environment'. The chapter argued that deep ecologists have been sociologically naïve, or perhaps worse, purposely ignorant, in making the assumption that all human populations across the globe are collectively and equally responsible for what they see as the current environmental crisis. Sociology has been concerned with the diverse and complex ways in which human societies are differentiated on grounds of caste, class, gender, ethnicity, age, sexuality and other forms of difference and inequality. Thus sociologists engaging with and developing green political and social theory have tended to be hostile to deep ecology particularly, and rightly so, in its earlier incarnations.

Social ecologists have a different conception of the causes of environmental crisis and argue that this is due to the forms of oppression and domination that characterize human societies, which enable and encourage the exploitation of the natural environment. I argued, however, that they almost go too far in the opposite direction in according no causal power to the non-human life-world. This is a position that might underplay the extent to which human society collectively exploits natural resources, and some eco-feminist and some eco-socialist approaches achieve a more balanced position, which combines theories of the human domination of the environment with an analysis of social division and difference.

Chapter 3 focused on the way environmentalist ideas have translated into policy, public discourse and political action. I suggested that environmentalism is a social movement with some coherence of philosophy and aim if very broadly defined. Sociologists have produced a range of explanations for social movement activity, some focusing on the motivations of individuals who join parties or engage in protests, others focusing on the social and economic changes that give rise to such movements. Eurocentrism is apparent in the failure of much social movement theory to account for the emergence and different formations of environmentalism in developing countries. In arguing that there is a specific variety of environmentalism, the 'environmentalism of the poor', some theorists have suggested that the differences between environmentalisms North and South is that the Southern movements are more diverse, and often defined by the specific and contingent circumstances of particular regions and localities. In addition, it has been suggested that Southern environmentalism is likely to be embedded in protest about intra-human forms of exclusion, deprivation and injustice. This is partly because environmental problems have a tendency to be more devastating on the flora and fauna of the Southern Hemisphere, including human communities.

Part of the difficulty here, in my view, is that current sociological approaches to environmental social movement organizations tend to be interested in either the social origins of social movements, or the ways movements manage to 'construct' environmental problems by framing public debate. I think this anthropocentric paradigm in social movement theory is also contributory to the Eurocentrism of many prevailing approaches. The 'environment' itself plays a role in defining what the issues are, and needs to be accounted for in theorizations of political protest. While issues are defined by social movement organizations, often through their influence on the media, the problems themselves may also be real. In developing countries, the proximity of many people to those problems, combined with social inequality, may be particularly significant for environmental activism.

Chapter 4 was based on the theme of historical change and focused on the 'transitions to modernity' in Europe, in which sociology, as it developed historically, was profoundly interested. Some of the issues of contemporary environmental concern focus on social and environmental impacts of the historical processes of urbanization and industrialization. These include such key issues as the causes and consequences of pollution, congestion, food safety, poor housing, ill health, loss of habitat and species diversity. Contemporary sociologists have developed some of the concepts utilized by classical sociologists, such as alienation, the division of labour, community, consumption and the representation of space and place in analysing contemporary formations of environment–society relations. Here, sociologists have a vital contribution to make. The theoretical, conceptual and methodological apparatus for the study of social difference and diversity, inclusion and exclusion, exploitation and oppression, can be deployed to avoid the over-general, gender-, class- and race-blind material that some deep ecologists have produced in the analysis of industrial consumer society.

While the social and environmental changes accompanying the transitions to modernity in Europe were historically and culturally specific, sociologists of development, postcolonialism and globalization have understood that they have had an impact on much of the planet. The relations between rich and poor countries, the impact of development and underdevelopment on environment–society relations, and the uneven global distribution of environmental problems, are questions which sociologists are well placed to answer. The key theme of Chapter 5 was the often global effects of environmental problems and their uneven concentration, distribution, experience and causality across the regions of the globe. Environmental problems may be caused by social practices associated with the development of modern industrial and consumer societies. Globalization itself is a highly contested term, with widely differing perspectives on the nature of the processes of change, its historical development and long-term future. While international organizations such as the UN have attempted to discuss initiatives and develop global policy programmes with respect to the environment, different kinds of social, economic and political interests almost inevitably conflict in such attempts.

Chapter 6 investigated human relations with other animal species. It argued that we continually interact with animals in our everyday lives, and that this pattern of interaction is historically constant although specific forms and sites of human–animal interactions shift over time. In the last century, dominant conceptions of animals and common practices towards animals have altered significantly, and in the areas of human–animal interaction we looked at – entertainment, food and pet-keeping – change is evident. In the late modern West, there has been increased concern about the welfare of food animals, and about the implications for human health of intensive farming methods. More people are keeping different kinds of animals as 'pets', but in similar ways to the viewing of animals as popular entertainment, attitudes to animals have become more 'decentred'. We are now more aware of and sympathetic towards the needs of different animal species. Even when we consider the use of animals in hunting, we can see not only a different conception of human–animal relations, but evidence of the increased influence of arguably more benign attitudes.

For me, one of the most interesting observations of the sociology of human–animal relations is the idea of co-dependency between humans and animals. This applies to the environment more broadly, and sociologists with very different theoretical and political perspectives have concurred that the environment is embedded in social relations. The title of this book suggests separation: *Environment* and *Society*, and at the end of writing it, I have come to feel I don't much like it! I hope the book indicates how important sociology can and will be, and indeed, has already been, in understanding the 'environment'. I hope also, however, that it will help question the very notion of 'environment' and 'society' as distinct. Both environment and society are co-dependent, and so very often the social and the 'natural' are constituted in complex and intimate relationships.

 **Glossary**

agency   action, referring to the ability (usually) of individuals and groups to change or alter their circumstances. Some green thinkers argue that non-humans have agency, particularly higher mammals.

alienation   separation, estrangement, that is unnatural/undesirable, socially caused. In Marxism 'alienation' is used to mean separation from oneself, particularly workers from their own labour. It has been used to describe separation from nature.

anomie   first developed by Emile Durkheim. Used to refer to a situation in which norms and values are disrupted and people experience social dislocation.

anthropocentric/anthropocentrism   a way of thinking or acting which is human-centred or primarily concerned with humans.

biocentric/biocentricity   a way of thinking or acting which is primarily concerned with the biosphere, or the planet and its variety of plant and animal species.

biodiversity   biological diversity of plant and animal life.

bioregionalism   first developed by the environmentalist Kirkpatrick Sale. A mode of social organization based on bioregions which are defined geographically by soil type, geographic features such as river basins, etc. People live within the ecological limits of their local bioregion.

biotechnology   the application of biological knowledge to production in the material world. Used to refer to new forms of technology that have changed or developed biological life forms, such as genetically modified plants and seeds.

charismatic mega-fauna   the fauna of a region are its indigenous species of animal; mega-fauna are large animals. It has been argued that some large animals are socially constructed as charismatic, particularly attractive, and are often represented aesthetically in environmentalist appeals for habitat conservation. Species vary according to regions of the globe: in Britain it could be foxes, and in the past, wolves; in the United States, bears, pumas; in India, elephants, tigers. In the 1970s, whales became the first creatures publicly defined and promoted in this way, through 'Save the Whale' campaigns.

co-constructionism   a theoretical understanding of social life where the social and the natural worlds are seen as constructed together. Thus the social and the natural are not categories we can 'separate out' and consider one without the other.

corporate greenwashing   the use of environmental/green labelling of products and practices that are not environmentally friendly, usually by commercial business. For example, crushable plastic bottles, which take up less room in a landfill site and are marketed as producing 'less waste', are still made of plastic which is not biodegradable.

critical realism   realism is a theoretical understanding of social and biological life which holds that objects have their own properties and powers and causal mechanisms. These may not always be apparent, but have effects, for example, viruses and diseases. In social theory it has meant an understanding of society and nature as having relatively autonomous (semi-independent) properties, and the acknowledgement that there are not always readily apparent structures which shape and constitute social and biological life. Critical realism argues that these properties and powers are also socially mediated, shaped and represented in certain ways.

cyborg   an amalgam of natural and artificial/artifactual objects. A whole range of objects could be described by this term: electronic cyberpets, the identity someone may assume in an Internet chat room, someone whose heart has a replacement metal valve, or who has an artificial hip joint, a genetically modified tomato.

dependency   the concept of dependency was developed by 'dependency theory' in the 1960s and 1970s. The theory suggests that, in a range of countries, particularly in the Southern Hemisphere or 'Third World' there is a limited ability to control key aspects of political and economic life because of the dominance of wealthy Northern economies in global trade.

development   the increasing satisfaction of needs, both economic and social. This has usually referred to economic growth on the capitalist and industrial model, and increased provision of social goods and services.

dialectical   used to analyse opposite phenomena which operate in interrelation. Change in one phenomenon causes change in the other and change takes place constantly. Society and the environment can be seen to be in a dialectical relationship.

disassembly   used to describe the taking apart of objects in the manufacturing process, such as animals on a slaughter line. The opposite of assembly, such as the assembly line of factory production.

discourses   sets of interrelated ideas about particular objects or phenomena which shape how those objects etc. are perceived and how we interact with them. Discourses are seen by Michel Foucault to shape social life and to carry relations of power.

division of labour   the specialization of work tasks through which different kinds of tasks or occupations are combined in a system of economic production.

dominant paradigm   the dominant way of thinking about things which shaped all forms of academic knowledge, social ideas and beliefs and modes of behaviour.

eco-system   an ecological community of interrelating parts including soils, water, plants, various species of animals. Considered as a unit.

(the) Enlightenment   the radical series of changes in philosophy, political and social thought in America and Europe in the eighteenth century, which stressed rationalism, technological progress, etc.

environmental justice   campaigns and arguments for environmental justice argue that environmental problems are unevenly distributed, and that certain powerful countries, regions, social groups are able to subject more vulnerable societies and groups to environmental hazards.

epistemology/epistemological   a theory of the nature of knowledge, or of how we know what we know. Realism, for example, argues that there are hidden influences in society that are not always observable, whereas positivism thinks that knowledge is factual when it is directly perceived and we can directly observe social facts. Interpretivists ague that we can only understand social life by investigating the meanings which people give to processes, experiences, phenomena.

essentialist/essentialism   the idea that objects have essential (fixed, unchanging) properties which can be seen cross-culturally and across historical time. It has been applied critically to certain social and political theories (feminism, Marxism, ecologism) which see relationships of hierarchy or exploitation as a feature of social development.

Eurocentric/Eurocentrism   a mode of thought or political behaviour which is centred on European knowledge, understanding and experience, and which ignores the differing experiences and knowledges from outside the European context.

exotic   having the charm of unfamiliarity, usually used to describe things that are 'foreign', from another part of the world. A Eurocentric concept, in which peoples and animals from the 'non-West' are defined as exotic.

exploitation   relationships of economic inequality in which one individual, organization or group is able to benefit from the work, labour or resources of another.

globalization   describes a series of economic, political and cultural changes and developments which draw parts of the globe together in closer relationships, creating for example, a single market, a common culture.

hybrid/hybridity   an object that is the result of the breeding of genetically dissimilar stock of plants or animals. In sociology, hybridity has been used to discuss the mixed composition, nature and origin of aspects of social life, particularly the interface between the social and the natural.

imperialism   used to describe, often critically, the economic, social and political relations between a colonized country and its colonizing power, and often the extraction of resources (labour, natural resources, etc.) by Western societies through inequitable trade agreements and other economic relations. The term also implies that imperialist powers attempt to continue and often increase the benefits they obtain from such relations.

intrinsic value   something which has value in itself rather than being valued for its use for other things or purposes.

liberal democratic/liberal democracy   a system of government based on elected parliamentary institutions, a 'free' press, the assumption of a free market economy and individual civil liberties (voting rights, freedom of speech and assembly, etc.)

lifestyle politics   often used critically to describe the adoption of certain changes in lifestyle as an alternative or as a part of the attempt to secure political and social change. While joining an environmental group is political, being a vegetarian or using traditional Chinese rather than Western medicine for example, is seen as a lifestyle choice, but also one which is critical of the status quo.

modernism/modernity   'modernity' is used to describe the result of the process of modernization (see below) and refers to the experience of European development. Modernism is used to describe the ways of thinking associated with rationalism and scientific knowledge, and the social and economic structures of industrial production. In sociology, modernism is used to describe approaches that use and develop the sociological concepts and theories that emerged in the nineteenth century, such as Marxist theories of capitalism and Weberian theories of rationalism.

modernization   sometimes another word for development. It is used to describe a particular path of development (social and economic change) which follows that of European societies in the eighteenth and nineteenth centuries and involves industrialization, urbanization, secularization and often also the development of a liberal democratic political system.

monoculture   a single culture. Can apply to societies that are standardized and lack diversity or the acceptance of diversity, or to natural/biological objects that have limited genetic diversity.

narratives   stories, ways of talking about, describing a phenomenon. Interpretivist sociologists argue that the study of people's narratives and how they interpret events and phenomena is the best way of capturing the complexity of social life.

new-ageism   a range of groups and beliefs often associated with Western paganism (Earth worship) which sees the world as entering a new era involving a radical and positive change in human relationships and priorities.

ontology/ontological   a theory of the nature, of being in the world, or 'how things are', such as Marxist theories of capitalism, feminist theories of patriarchy, and deep green theories of anthropocentrism.

oppression   an intense form of social domination (see below) in which structured relations of social power significantly restrict the agency (see above) of oppressed groups.

patriarchy   literally the 'rule of the fathers'. Used by some feminists to describe and analyse a social system of structured relationships, which exclude, exploit and oppress women. Not all men benefit from this system, and women may be agents of patriarchal power and domination.

post-industrial/post-industrialism   the idea that developed industrial societies have moved to post-industrial production involving services rather than manufactures. Also the notion that such societies have a critical or reflexive attitude to the process and structures of industrialism.

postmodern/postmodernism/postmodernity   postmodernism questions the basis of modernism and the theories that attempted to discern general underlying principles about social life. Rather than generate all-embracing theories or grand narratives, postmodernism asserts we should focus on the production of more micro 'situated' knowledge. Many theorists argue that society is also now postmodern, as the structures and processes of modernity have been disrupted, challenged and undermined, and social life and social theory are now characterized by uncertainty.

sentience/sentiency   the ability to sense certain phenomena, particularly the ability to experience pain, suffering or pleasure.

social constructionism   an approach which holds in stronger or weaker form that there is no 'objective' reality which has not been constructed through narrative or discourse (see both terms, above).

social difference   the ideas, perspectives and experiences of those from various social locations of age, class, sex, sexuality, race, ethnicity, regional context, etc. are all forms and expressions of social difference.

social domination   the structured relations of social power so that certain groups are in positions of strategic advantage in relation to others.

social exclusion   can take strong or weak forms and involves certain groups in society being relatively excluded from economic and social benefits and positions of power and authority, such as educational achievement and political representation.

social inequality   unequal power, status, outcome, distribution of goods between different groups in society.

social stratification   structured social inequalities between different groupings of people. Like strata in rock, societies are seen to stratify hierarchically.

species-being   the nature of a particular species, its mode of life shaped by its biology and sociality. Used by Marx to discuss human nature.

structure   an organizing principle or mechanism of social life which shapes social behaviour in important ways. Social institutions such as workplaces, schools and the household can be seen as social structures, as can the practices within them, such as work, marriage, education, etc.

technocratic   used to describe approaches which see environmental problems as 'technical' issues which can be resolved with application of scientific knowledge and technical innovation.

underdevelopment   see development. Development theorists argued that richer Western societies underdeveloped poorer counties by exploitative trade relations.

vegan/veganism   someone who avoids using animal products. In addition to not eating meat or fish, a vegan does not eat eggs, dairy products or honey, and does not wear leather, fur, wool or silk. Like vegetarianism (below) there are degrees of stringency among those adopting such practices.

vegetarian/vegetarianism   someone who avoids eating animal flesh (meat, fish), products made with animal derivatives (such as gelatin from bones) and perhaps also animal products such as eggs produced in intensive conditions.

wilderness   land which is relatively unmodified by human settlement and patterns of life.

zoological gaze   the practice of looking at other animal species, and the desirability of such observation. This can be seen in wildlife documentaries and viewing animals in aquaria, safari parks and zoos.

 **Bibliography**

Adams, C.J. (1976) 'The Inedible Complex: The Political Implications of Vegetarianism', *Second Wave* 4, 1: 36–42.

Adams, C.J. (1990) *The Sexual Politics of Meat*. Cambridge: Polity.

Adams, C.J. (1994) *Neither Man nor Beast: Feminism and the Defense of Animals*. New York: Continuum.

Adams, C.J. (1995) 'Women Battering and Harm to Animals', in C.J. Adams and J. Donovan (eds) *Animals and Women: Feminist Theoretical Explanations*. London: Duke University Press.

Adams, C.J. and Donovan, J. (eds) (1995) *Animals and Women: Feminist Theoretical Explanations*. London: Duke University Press.

Agarwal, B. (1992) 'The Gender and Environment Debate: Lessons from India', *Feminist Studies* 18, 1: 119–58.

Albrow, M. (1996) *The Global Age*. Cambridge: Polity.

Alcoff, L. and Potter, E. (eds) (1993) *Feminist Epistemologies*. London: Routledge.

Almond, G. (1970) *Political Development*. Boston, MA: Little Brown.

Almond, B. (1995) 'Rights and Justice in the Environment Debate', in D. Cooper and J. Palmer (eds) *Just Environments: Intergenerational, International and Interspecies Issues*. London: Routledge.

Almond, G. and Coleman, J. (eds) (1965) *The Politics of Developing Areas*. Princeton, NJ: Princeton University Press.

Almond, G. and Powell, B. (1966) *Comparative Politics: A Developmental Approach*. Boston, MA: Little Brown.

Amin, S. (1997) *Capitalism in the Age of Globalization*. London: Zed.

Anderson, B. (1991) *Imagined Communities: Reflections on the Origin and Spread of Nationalism*. London: Verso.

Antonio, D. (1995) 'Of Wolves and Women', in C.J. Adams and J. Donovan (eds), *Animals and Women: Feminist Theoretical Explanations*. London: Duke University Press.

Appadurai, A. (1990) 'Disjuncture and Difference in the Global Cultural Economy', in M. Featherstone (ed.) *Global Culture: Nationalism, Globalization, Modernity*. London: Sage.

Attfield, R. (1983) *The Ethics of Environmental Concern*. Oxford: Blackwell.

Bahro, R. (1982) *Socialism and Survival*. London: Heretic Books.

Bahro, R. (1984) *From Red to Green: Interview with New Left Review*. London: Verso.

Bahro, R. (1986) *Building the Green Movement*. London: Heretic Books.

Baran, P. (1957) *The Political Economy of Growth*. New York: Monthly Review Press.

Barthes, R. (1972) *Mythologies*. New York: Hill and Wang.

Barthes, R. (1979) 'Towards a Psychosociology of Contemporary Food', in R. Forster and O. Ranum (eds) *Food and Drink in History*. Baltimore, MD: Johns Hopkins University Press.

Barry, J. (1999) *Environment and Social Theory*. London: Routledge.

Baudrillard, J. (1983) *Simulations*. New York: Semiotext(e).

Bauman, Z. (1991) *Modernity and Ambivalence*. Cambridge: Polity.

Beck, U. (1992) *The Risk Society*. London: Sage.

Beck, U. (1995) *Ecological Politics in an Age of Risk*. Cambridge: Polity.

Beck, U. (1996) 'Risk Society and the Provident State', in S. Lash, B. Szerszynski and B. Wynne (eds) *Risk, Environment and Modernity: Towards a New Ecology*. London: Sage.

Beck, U. (1999) *World Risk Society*. Cambridge: Polity.

Beder, S. (1997) *Global Spin: The Corporate Assault on Environmentalism*. Dartington: Green Books.

Bell, M.M. (1998) *An Invitation to Environmental Sociology*. Thousand Oaks, CA/London: Pine Forge Press/Sage.

Bendix, R. (1967) 'Tradition and Modernity Reconsidered', *Studies in Comparative Society and History* 9, 3: 141–70.

Benton, T. (1985) 'Realism and Social Science', in R. Edgley and R. Osborne (eds) *Radical Philosophy Reader*. London: Verso.

Benton, T. (1989) 'Marxism and Natural Limits: An Ecological Critique and Reconstruction', *New Left Review* 178: 51–86.

Benton, T. (1991) 'Biology and Social Science: Why the Return of the Repressed Should be Given a (Cautious) Welcome', *Sociology* 25: 1–29.

Benton, T. (1993) *Natural Relations: Ecology, Animal Rights and Social Justice*. London: Verso.

Benton, T. (1994) 'Biology and Social Theory in the Environment Debate', in M. Redclift and T. Benton (eds) *Social Theory and the Global Environment*. London: Routledge.

Benton, T. (ed.) (1996) *The Greening of Marxism*. London: Guilford.

Benton, T. (1998) 'Why are Sociologists Nature-phobes?' paper to the Centre for Critical Realism Conference – After Postmodernism: Critical Realism?, University of Essex.

Berger, P. (1987) *The Capitalist Revolution*. Aldershot, Hants: Wildwood House.

Bhaskar, R. (1978) *A Realist Theory of Science*. Sussex: Harvester.

Bhaskar, R. (1989) *The Possibility of Naturalism*. Hemel Hempstead: Harvester Wheatsheaf.

Biehl, J. (1988) 'What is Social Ecofeminism?', *Green Perspectives* 11: 1–8.

Biehl, J. (1991) *Finding Our Way: Rethinking Ecofeminist Politics*. Montreal: Black Rose Books.

Birke, L. (1986) *Women, Feminism and Biology*. Brighton: Harvester.

Bloor, D. (1999) 'Anti-Latour', *Studies in the History and Philosophy of Science* 30: 81–112.

Boerner, C. and Lambert, T. (1995) 'Environmental Injustice', *The Public Interest* 95, 118: 61–82.

Boggs, C. (1986) 'Social Movements and Political Power', in *Emerging Forms of Political Radicalism in the West*. Philadelphia, PA: Temple University Press.

Bookchin, M. (1971) *Post-Scarcity Anarchism*. Berkeley, CA: Ramparts Press.

Bookchin, M. (1980) *Towards an Ecological Society*. Montreal: Black Rose Books.

Bookchin, M. (1986) *The Modern Crisis*. Philadelphia, PA: New Society.

Bookchin, M. (1989) *Remaking Society*. Montreal: Black Rose Books.

Bookchin, M. (1990) *The Philosophy of Social Ecology*. Montreal: Black Rose Books.

Bookchin, M. (1991) *The Ecology of Freedom: The Emergence and Dissolution of Hierarchy*. Montreal: Black Rose Books.

Bookchin, M. (1995) *Re-enchanting Humanity*. London: Cassell.

Bordo, S. (1987) *The Flight to Objectivity: Essays on Cartesianism and Culture*. Albany: SUNY Press.

Bostock, S. (1993) *Zoos and Animal Rights*. London: Routledge.

Bourdieu, P. (1984) *Distinction*. London: Routledge and Kegan Paul.

Bradford, G. (1989) *How Deep is Deep Ecology?* Ojai: Times Change Press.

Braun, B. and Castree, N. (eds) (1998) *Remaking Reality: Nature at the Millennium*. London: Routledge.

Brown, D. (1970) *Bury my Heart at Wounded Knee*. London: Vintage.

Bryant, B. (1996) *Twyford Down: Roads, Campaigning and Environmental Law*. London: E. and F.N. Spon.

Bullard, R. (1990) *Dumping in Dixie: Race, Class and Environmental Quality*. Boulder, CO: Westview Press.

Bullard, R. (1993) *Confronting Environmental Racism: Voices from the Grassroots*. Boston, MA: South End Press.

Bunyard, P. and Morgan-Grenville, F. (eds) (1987) *The Green Alternative*. London: Methuen.

Bürklin, W.P. (1988) 'A Politico-Economic Model Instead of a Sour Grapes Logic: A Reply to Herbert Kitschell's Critique', *European Sociological Review* 4, 2: 161–6.

Burningham, K. and Cooper, G. (1999) 'Being Constructive: Social Constructionism and the Environment', *Sociology* 33, 2: 297–316.

Butler, T. and Savage, M. (eds) (1995) *Social Change and the Middle Classes*. London: UCL Press.

Buttel, F. (1997) 'Social Institutions and Environmental Change', in M. Redclift and G. Woodgate (eds) *The International Handbook of Environmental Sociology*. Cheltenham: Edward Elgar.

Byrne, P. (1997) *Social Movements in Britain*. London: Routledge.

Capra, F. (1976) *The Tao of the New Physics*. London: Flamingo.

Capra, F. (1983) *The Turning Point*. London: Flamingo.

Callinicos, A. *et al.* (1994) *Marxism and the New Imperialism*. London: Bookmarks.

Castells, M. (1996) *The Rise of the Network Society*. Oxford: Blackwell.

Castells, M. (1998) *The End of the Millennium*. Oxford: Blackwell.

Catton, W.R. and Dunlap, R.E. (1978) 'Environmental Sociology: A New Paradigm', *The American Sociologist* 13: 41–9.

Catton, W.R. and Dunlap, R.E. (1980) 'A New Paradigm for Post-Exuberant Sociology', *American Behavioral Scientist* 24, 1: 15–47.

Charles, N. and Kerr, M. (1988) *Women, Food and Families*. Manchester: Manchester University Press.

Chase, B. (ed.) (1991) *Defending the Earth: A Dialogue Between Murray Bookchin and Dave Foreman*. Boston, MA: South End Press.

Chatterjee, P. and Finger, M. (1994) *The Earth Brokers: Power, Politics and World Development*. London: Routledge.

Cheney, J. (1994) 'Nature/Theory/Difference: Ecofeminism and the Reconstruction of Environmental Ethics', in K. Warren (ed.) *Ecological Feminism*. London: Routledge.

Christ, C.P. (1992) 'Spiritual Quest and Women's Experience', in C.P. Christ and J. Plaskow (eds) *Womanspirit Rising*. New York: HarperCollins.

Christ, C.P. and Plaskow, J. (eds) (1992) *Womanspirit Rising*. New York: HarperCollins.

Clapham, C. (1985) *Third World Politics*. London: Croom Helm.

Collard, A. with Contrucci, J. (1988) *Rape of the Wild: Man's Violence Against Animals and the Earth*. London: The Women's Press.

Cole, H. *et al.* (eds) (1973) *Thinking About the Future: A Critique of the 'Limits to Growth'*. London: Chatto and Windus.

Collier, A. (1994) *Critical Realism*. London: Verso.

Collins, P. (1990) *Black Feminist Thought*. London: Routledge.

Commoner, B. (1972) *The Closing Circle: Confronting the Environmental Crisis*. New York: Bantam.

Comninou, M. (1995) 'Speech, Pornography and Hunting', in C.J. Adams and J. Donovan (eds) *Animals and Women: Feminist Theoretical Explanations*. London: Duke University Press.

Connelly, J. and Smith, G. (1999) *Politics and the Environment: From Theory to Practice*. London: Routledge.

Cooper, D. (1993) 'Human Sentiment and the Future of Wildlife', *Environmental Values* 2: 335–46.

Cooper, D. (1995) 'Other Species and Moral Reason', in D. Cooper and J. Palmer *Just Environments: Intergenerational, International and Interspecies Issues*. London: Routledge.

Cooper, D. and Palmer, J. (eds) (1995) *Just Environments: Intergenerational, International and Interspecies Issues*. London: Routledge.

Corea, G. (1985) *The Mother Machine*. London: The Women's Press.

Cotgrove, S. (1982) *Catastrophe or Cornucopia: The Environment, Politics, and the Future*. Chichester: John Wiley and Sons.

Cotgrove, S. (1991) 'Sociology and the Environment: Cotgrove Replies to Newby', *Network* 51, October: 5.

Craib, I. (1992) *Modern Social Theory: An Introduction*. Hemel Hempstead: Harvester Wheatsheaf.

Cudworth, E. (1998) 'Gender, Nature and Domination', unpublished PhD thesis, University of Leeds.

Cudworth, E. (1999) 'The Structure/Agency Debate in Environmental Sociology: Towards a Structural and Realist Approach', *Social Politics Papers*, No. 4, University of East London.

Cuomo, C. (1995) 'Ecofeminism, Deep Ecology and Human Population', in K. Warren (ed.) *Ecological Feminism*. London: Routledge.

Daly, M. (1984) *Pure Lust*. London: The Women's Press.

Daly, M. (1986) *Beyond God the Father*.1986 edition, London: The Women's Press. (First published 1973.)

Daly, M. (1979) *Gyn/Ecology: The Metaethics of Radical Feminism*. London: The Women's Press.

Davies, J. (1962) 'Towards a Theory of Revolution', *American Sociological Review* 27.

Davion, V. (1994) 'Is Ecofeminism Feminist?', in K. Warren (ed.) *Ecological Feminism*. London: Routledge.

Della Porta, D. and Diani, M. (1999) *Social Movements: An Introduction*. Oxford: Blackwell.

Devall, B. (1990) *Simple Means, Rich in Ends*. London: Green Print.

Devall, B. and Sessions, G. (1985) *Deep Ecology: Living as if Nature Mattered*. Layton: Gibbs M. Smith.

d'Eaubonne, F. (1980) 'Le feminisme ou la mort', in E. Marks and I. de Courtivron (eds) *New French Feminisms: An Anthology*. Amherst: University of Massachusetts Press.

Diamond, I. and Orenstein, G. (eds) (1990) *Reweaving the World*. San Francisco, CA: Sierra Club Books.

Dickens, P. (1992) *Society and Nature: Towards a Green Social Theory*. London: Harvester Wheatsheaf.

Dickens, P. (1996) *Reconstructing Nature: Alienation, Emancipation and the Division of Labour*. London: Routledge.

Dickens, P. (2000) *Social Darwinism*. Buckingham: Open University Press.

Dickens, P. (2001) 'Linking the Social and Natural Sciences: Is Capital Modifying Human Biology in Its Own Image? *Sociology* 35, 1: 93–110.

Dobson, A. (1990) *Green Political Thought*. London: Routledge.

Dobson, A. (ed.) (1991) *The Green Reader*. London: André Deutsch.

Donovan, J. (1990) 'Animal Rights and Feminist Theory', *Signs* 15, 2: 350–75.

Donovan, J. and Adams, C.J. (eds) (1996) *Beyond Animal Rights: A Feminist Ethic for the Treatment of Animals*. New York: Continuum.

Doubiago, S. (1989) 'Mama Coyote Talks to the Boys', in J. Plant (ed.) *Healing the Wounds: The Promise of Ecofeminism*. London: Green Print.

Doyle, T. and McEachern, D. (1998) *Environment and Politics*. London: Routledge.

Dunlap, R.E. and Catton, W.R. (1993) 'The Development, Current Status and Probable Future of Environmental Sociology: Toward an Ecological Sociology', *Annals of the International Institute of Sociology* 3: 263–84.

Durkheim, E. (1935) *The Division of Labour in Society*. New York: Free Press.

Durning, A.T. and Brough, H.B. (1995) 'Animal Farming and the Environment', in D. Cooper and J. Palmer (eds) *Just Environments: Intergenerational, International and Interspecies Issues*. London: Routledge.

Eastlea, B. (1981) *Science and Sexual Oppression: Patriarchy's Confrontation with Women and Nature*. London: Weidenfeld and Nicolson.

Eckersley, R. (1989) 'Green Politics and the New Class: Selfishness or Virtue?' *Political Studies* 37: 205–23.

Eckersley, R. (1992) *Environmentalism and Political Theory*. London: UCL Press.

Eder, K. (1996) *The Social Construction of Nature*. London: Sage.

Eisenstadt, S. (1966) *Modernization: Protest and Change*. Englewood Cliffs, NJ: Prentice Hall.

Eisler, R. (1990) *The Chalice and the Blade*. London: Unwin Hyman. (First published 1987.)

Ekins, P. (ed.) (1986) *The Living Economy: A New Economics in the Making*. London: Routledge and Kegan Paul.

Ekins, P. (1992) *A New World Order: Grassroots Movements for Global Change*. London: Routledge.

Elias, N. (1994) *The Civilizing Process*. Oxford: Blackwell.

Elkington, J. and Burke, T. (1987) *The Green Capitalists: Industry's Search for Environmental Excellence*. London: Gollancz.

Elkington, J. and Hailes, J. (1988) *The Green Consumer Guide: From Shampoo to Champagne: High Street Shopping for a Better Environment.* London: Gollancz.

Elliot, R. and Gare, A. (eds) *Environmental Philosophy.* Milton Keynes: Open University Press.

Eyerman, R. and Jamison, A. (1991) *Social Movements: A Cognitive Approach.* Cambridge: Polity Press.

Faludi, S. (1992) *Backlash.* London: Chatto and Windus.

Featherstone, M. (1988) 'In Pursuit of the Postmodern: An Introduction', *Theory, Culture & Society* 5, 2–3: 195–215.

Featherstone, M. (ed.) (1990) *Global Culture: Nationalism, Globalization, Modernity.* London: Sage.

Ferguson, K. (1993) *The Man Question.* Berkeley: University of California Press.

Flax, J. (1990) 'Postmodernism and Gender Relations in Feminist Theory', in L. Nicholson (ed.) *Feminism/Postmodernism.* London: Routledge.

Foreman, D. and Haywood, B. (eds) (1989) *Ecodefense: A Field Guide to Monkeywrenching.* Tucson, AZ: Ned Ludd Books.

Fox, W. (1984) 'Deep Ecology: A New Philosophy of Our Time?' *The Ecologist* 14: 5–6.

Fox, W. (1989) 'The Deep Ecology–Ecofeminism Debate and its Parallels', *Environmental Ethics* 11: 5–25.

Fox, W. (1990) *Towards a Transpersonal Ecology.* Boston, MA: Shambhala.

Frank, A.G. (1969) *Capitalism and Underdevelopment in Latin America: Historical Studies of Chile and Brazil.* New York: Monthly Review Press.

Frank, A.G. (1971) *The Sociology of Development and the Underdevelopment of Sociology.* London: Pluto Press.

Frankland, E.G. and Schoonmaker, D. (1992) *Between Protest and Power: The Green Party in Germany.* Boulder, CO: Westview Press.

Franklin, A. (1999) *Animals and Modern Cultures: A Sociology of Human–Animal Relations in Modernity.* London: Sage.

Frey, R.G. (1980) *Interests and Rights: The Case Against Animals.* Oxford: Clarendon.

Frey, R.G. (1983) *Rights, Killing and Suffering.* Oxford: Blackwell.

Friedan, B. (1965) *The Feminine Mystique.* Harmondsworth: Penguin.

Frisby, D. and Featherstone, M. (eds) (1997) *Simmel on Culture.* London: Sage.

Fukuyama, F. (1992) *The End of History and the Last Man.* London: Hamish Hamilton.

Fuss, D. (1989) *Essentially Speaking.* London: Routledge.

Gans, M. (1962) *The Urban Villagers.* New York: Free Press of Glencoe.

Gare, A. (1996) 'Soviet Environmentalism: The Path Not Taken', in T. Benton (ed.) *The Greening of Marxism.* New York: Guilford Press.

Garner, R. (1993a) *Animals, Politics and Morality.* Manchester: Manchester University Press.

Garner, R. (1993b) 'Political Animals', *Parliamentary Affairs*: 333–52.

George, S. (1988) *A Fate Worse than Debt*. Harmondsworth: Penguin.

George, S. (1990) *Ill Fares the Land*. Harmondsworth: Penguin.

Giddens, A. (1971) *Capitalism and Modern Social Theory*. Cambridge: Cambridge University Press.

Giddens, A. (1973) *The Class Structure of Advanced Societies*. London: Hutchinson.

Giddens, A. (1984) *The Constitution of Society*. Cambridge: Polity.

Giddens, A. (1987) *Social Theory and Modern Sociology*. Stanford, CA: Stanford University Press.

Giddens, A. (1990) *The Consequences of Modernity*. Cambridge: Polity.

Giddens, A. (1991) *Modernity and Self-identity*. Cambridge: Polity.

Gilpin, R. (1987) *The Political Economy of International Relations*. Princeton, NJ: Princeton University Press.

Gimbutas, M. (1982) *The Goddesses and Gods of Old Europe*. Berkeley: University of California Press.

Girouard, M. (1985) *Cities and People*. London: Yale University Press.

Goldblatt, D. (1996) *Social Theory and the Environment*. Cambridge: Polity Press.

Goldsmith, E. (1972) *A Blueprint for Survival*. London: Tom Stacey.

Goodin, R. (1992) *Green Political Theory*. Cambridge: Polity.

Gorz, A. (1980) *Ecology as Politics*. London: Pluto.

Gorz, A. (1982) *Farewell to the Working Class: An Essay in Post-Industrial Socialism*. London: Pluto.

Gorz, A. (1985) *Paths to Paradise: On the Liberation from Work*. London: Pluto.

Gorz, A. (1994) *Capitalism, Socialism, Ecology*. London: Verso.

Grant, W. (1993) 'Transnational Companies and Environmental Policy Making: The Trend of Globalization', in J.D. Lieferink, P.D. Lowe and A.P.J. Mol (eds) *European Integration and Environmental Policy*. London: Belhaven.

Greer, G. (1985) *Sex and Destiny: The Politics of Human Fertility*. London: Picador.

Greider, W. (1997) *One World, Ready or Not: The Manic Logic of Global Capitalism*. New York: Simon and Schuster.

Griffin, S. (1983) Preface to L. Caldecott and S. Leyland (eds) *Reclaim the Earth*. London: The Women's Press.

Griffin, S. (1984) *Woman and Nature*. London: The Women's Press.

Griffin, S. (1994) *A Chorus of Stones*. London: The Women's Press.

Guha, R. (1997) 'The Environmentalism of the Poor', in R. Guha and J. Martinez-Alier (eds) *Varieties of Environmentalism*. London: Earthscan.

Hall, S. and Gieben, B. (eds) (1992) *Formations of Modernity*. Cambridge: Polity/Oxford University Press.

Hall, S. and Jacques, M. (eds) (1989) *New Times: The Changing Face of Politics in the 1990s*. London: Lawrence and Wishart in association with *Marxism Today*.

Hall, S. *et al.* (1978) *Policing the Crisis*. London: Macmillan.

Hannigan, J. (1995) *Environmental Sociology: A Social Constructionist Perspective*. London: Routledge.

Haraway, D. (1989) *Primate Visions: Gender, Race and Nature in the World of Modern Science*. London: Routledge.

Haraway, D. (1991) *Simians, Cyborgs and Women: The Reinvention of Nature*. London: Free Association Press.

Haraway, D. (1997) *Modest_Witness@Second_Millennium. FemaleMan_Meets_OncoMouse*. London: Routledge.

Hardin, G. (1977) 'The Tragedy of the Commons', in G. Hardin and J. Baden (eds) *Managing the Commons*. San Francisco, CA: W.H. Freeman and Co.

Harding, S. (1991) *Whose Science? Whose Knowledge?* Milton Keynes: Open University Press.

Harding, S (1993) 'Rethinking Standpoint Epistemology: What is Strong Objectivity?', in L. Alcoff and E. Potter (eds) *Feminist Epistemologies*. London: Routledge.

Harrison, D. (1988) *The Sociology of Modernization and Development*. London: Unwin Hyman.

Hartman, B. (1987) *Reproductive Rights and Wrongs: The Global Politics of Reproduction Control and Contraceptive Choice*. New York: Harper and Row.

Hartsock, N. (1983) *Money, Sex and Power*. Boston, MA: Northeastern University Press.

Hartsock, N. (1987) 'The Feminist Standpoint: Developing the Ground for a Specifically Feminist Historical Materialism', in S. Harding (ed.) *Feminism and Methodology*. Milton Keynes: Open University Press.

Harvey, D. (1985) *Consciousness and the Urban Experience: Studies in the History and Theory of Capitalist Urbanization*. Oxford: Basil Blackwell.

Harvey, D. (1990) *The Condition of Postmodernity*. Cambridge: Polity.

Hayward, T. (1990) 'Ecosocialism – Utopian and Scientific', *Radical Philosophy* 56, Autumn.

Hekman, S. (1990) *Gender and Knowledge*. Cambridge: Polity.

Held, D. (1991) 'Democracy, the Nation-State and the Global System', in D. Held (ed.) *Political Theory Today*. Cambridge: Polity Press.

Held, D. (1995) *Democracy and the Global Order: From the Modern State to Cosmopolitan Governance*. Cambridge: Polity Press.

Held, D. (1996) *Models of Democracy*, 2nd edn. Cambridge: Polity Press.

Held, D., McGrew, A., Goldblatt, D. and Peraton, J. (1999) *Global Transformations*. Cambridge: Polity Press.

Hirst, P. (1997) 'The Global Economy: Myths and Realities', *International Affairs* 73.

Hirst, P. and Thompson, G. (1996) *Globalization in Question: The International Economy and the Possibilities of Governance*. Cambridge: Polity Press.

Hobsbawm, E. (1968) *Industry and Empire*. Harmondsworth: Penguin.

Holden, A. (2000) *Environment and Tourism*. London: Routledge.

Hulsberg, W. (1988) *The German Greens: A Social and Political Profile*. London: Verso.

Huntington, S.P. (1965) 'Political Development and Political Decay', *World Politics* 17, 3: 386–430.

Huntington, S.P. (1968) *Political Order in Changing Societies*. New Haven, CT: Yale University Press.

Huntington, S.P. (1991) *The Third Wave: Democratization in the Late Twentieth Century*. Norman: University of Oklahoma Press.

Huntington, S.P. (1996) *The Clash of Civilizations and the Remaking of the World Order*. New York: Simon and Schuster.

Inglehart, R. (1977) *The Silent Revolution: Changing Values and Political Styles among Western Publics*. Princeton, NJ: Princeton University Press.

Inglehart, R. (1990) *Culture Shift in Advanced Industrial Society*. Princeton, NJ: Princeton University Press.

Inglehart, R. (1995) 'Public Support for Environmental Protection: Objective Problems and Subjective Values in 43 Societies', *Political Science and Politics* 28, 1: 57–72.

Ingold, T. (1986) *The Appropriation of Nature*. Manchester: Manchester University Press.

Ingold, T. (1994) 'From Trust to Dominion: An Alternative History of Human–Animal Relations', in A. Manning and J. Serpell (eds) *Animals and Human Society*. London: Routledge.

Irvine, S. (1989) *Beyond Green Consumerism*. London: Friends of the Earth.

Irvine, S. and Ponton, A. (1988) *A Green Manifesto: Policies for a Green Future*. London: Macdonald Optima.

Irwin, A. (2001) *Sociology and the Environment*. Cambridge: Polity Press.

Jacobs, M. (1996) *The Politics of the Real World*. London: Earthscan.

Jain, S. (1991) 'Standing Up for Trees: Women's Role in the Chipko Movement', in S. Sontheimer (ed.) *Women and the Environment: A Reader*. London: Earthscan.

Jameson, F. (1984) 'Postmodernism or the Cultural Logic of Late Capitalism', *New Left Review* 146: 53–93.

Jamison, A., Eyerman, R. and Cramer, J. (1990) *The Making of the New Environmental Consciousness: A Comparative Study of the Environmental Movements in Sweden, Denmark and the Netherlands*. Edinburgh: Edinburgh University Press.

Johnson, A. (1995) 'Barriers to Fair Treatment of Non-human Life', in D. Cooper and J. Palmer (eds) *Just Environments: Intergenerational, International and Interspecies Issues*. London: Routledge.

Jones, M. and Wangari, M. (1983) 'Greening the Desert: Women of Kenya Reclaim the Land', in L. Caldecott and S. Leyland (eds) *Reclaim the Earth*. London: The Women's Press.

Kappeler, S. (1995) 'Speciesism, Racism, Nationalism . . . or the Power of Scientific Subjectivity', in C.J. Adams and J. Donovan (eds) *Animals and Women: Feminist Theoretical Explanations*. London: Duke University Press.

Keller, E.F. (1985) *Reflections on Gender and Science*. New Haven: Yale University Press.

Keller, E F. (1992) *Secrets of Life, Secrets of Death*. London: Routledge.

Kelly, P. (1984) *Fighting for Hope*. London: Chatto and Windus.

Kemp, P. and Wall, D. (1990) *A Green Manifesto for the 1990s*. Harmondsworth: Penguin.

Kerr, C., Dunlop, J.T., Harbison, F.H., and Myers, C. (1960) *Industrialism and Industrial Man*. London: Heinemann.

Kheel, M. (1995) 'License to Kill: An Eco-feminist Critique of Hunters' Discourse', in C.J. Adams and J. Donovan (eds) *Animals and Women: Feminist Theoretical Explanations*. London: Duke University Press.

King, Y. (1990) 'Healing the Wounds: Feminism, Ecology and Nature/Culture Dualism', in I. Diamond and G. Orenstein (eds) *Reweaving the World*. San Francisco, CA: Sierra Club Books.

Kitsuse, J.I. and Spector, M. (1981) 'The Labeling of Social Problems' in E. Rubington and M.S. Weinberg (eds) *The Study of Social problems: Five Perspectives*. New York: Oxford University Press.

Kropotkin, P. (1955) *Mutual Aid*. New York: Extending Horizons Books.

Krugman, P. (1996) *Pop Internationalism*. Boston, MA: MIT Press.

Lappé, F.M. and Collins, J. (1978) *Food First: Beyond the Myth of Scarcity*. London: Abacus.

Lash, S. (1990) *The Sociology of Postmodernism*. London: Routledge.

Lash, S. and Urry, J. (1987) *The End of Organized Capitalism*. Cambridge: Polity Press.

Lash, S. and Urry, J. (1994) *Economies of Signs and Spaces*. London: Sage.

Latouche, S. (1991) *In the Wake of the Affluent Society*. London: Zed.

Latour , B. (1987) *Science in Action*. Milton Keynes: Open University Press.

Latour, B. (1993) *We Have Never Been Modern*. Hemel Hempstead: Harvester Wheatsheaf.

Latour, B. (1999) *Pandora's Hope: Essays on the Reality of Science Studies*. London: Harvard University Press.

Layder, D. (1994) *Understanding Social Theory*. London: Sage.

Lee, D. and Newby. H. (1991) *The Problem of Sociology: An Introduction to the Discipline*. London: Hutchinson.

Lefebvre, H. (1991) *The Production of Space*. Oxford: Blackwell.

Leff, E. (1996) 'Marxism and the Environmental Question: From the Critical Theory of Production to an Environmental Rationality for Sustainable Development', in T. Benton (ed.) *The Greening of Marxism*. New York: Guilford Press.

Leyland, S. (1983) 'Feminism and Ecology: Theoretical Considerations', in L. Caldecott and S. Leyland, *Reclaim the Earth*. London: The Women's Press.

Leys, C. (ed.) (1969) *Political Change in Developing Societies*. Cambridge: Cambridge University Press.

Light, A. (ed.) (1998) *Social Ecology after Bookchin*. New York: Guilford Press.

Lovelock, J. (1979) *Gaia: A New Look at Life on Earth*. Oxford: Oxford University Press.

Lovelock, J. (1989) *The Ages of Gaia: A Biography of our Living Earth*. Oxford: Oxford University Press.

Lovenduski, J. and Randall, V. (1993) *Contemporary Feminist Politics*. Oxford: Oxford University Press.

Lowe, P. and Rüdig, W. (1986) 'Review Article: Political Ecology and the Social Sciences', *British Journal of Political Science* 16: 513–50.

Lukes, S. (1973) *Emile Durkheim*. London: Allen Lane.

Lupton, D. (1996) *Food, the Body and the Self*. London: Sage.

Lyotard, J.-F. (1984) *The Postmodern Condition: A Report on Knowledge*. Minneapolis: University of Minnesota Press.

McCarthy, J.D. and Zald, M.N. (eds) (1979) *The Dynamics of Social Movements: Resource Mobilization, Social Control and Tactics*. Cambridge, MA.: Winthrop.

McCarthy, J.D. and Zald, M.N. (1987) 'The Trend of Social Movements in America: Professionalization and Resource Mobilization', in M.N. Zald and J.D. McCarthy, *Social Movements in an Organizational Society*. New Brunswick: Transaction.

McGrew, A.G. (ed.) (1997) *The Transformation of Democracy? Globalization and Territorial Democracy*. Cambridge: Polity Press.

McCormick, J. (1991) *British Politics and the Environment*. London: Earthscan.

Macnaghten, P. and Urry, J. (1995) 'Towards a Sociology of Nature', *Sociology* 29, 2: 203–20.

Macnaghten, P. and Urry, J. (1998) *Contested Natures*. London: Sage.

Manes, C. (1990) *Green Rage*. Boston, MA: Little Brown.

Mann, M. (1997) 'Has Globalization Ended the Rise and Rise of the Nation State?' *Review of International Political Economy* 4.

Marshall, D. (1993) *Demanding the Impossible: A History of Anarchism*. London: Macmillan.

Martell, L. (1994) *Ecology and Society: An Introduction*. Cambridge: Polity.

Marx, K. (1975) *Early Writings*, edited by L. Colletti. Harmondsworth: Penguin.

Marx, K. (1976) *Capital*. Harmondsworth: Penguin.

Marx, K. and Engels, F. (1983) *Manifesto of the Communist Party*. Harmondsworth: Penguin.

Mathias, P. (1969) *The First Industrial Nation*. London: Methuen.

Meadows, D., Randers, H.J. and Behrens, W.W. (1972) *The Limits to Growth*. New York: Universe Books.

Meadows, D.H., Meadows, D.L. and Randers, J. (1992) *Beyond the Limits: Global Collapse or a Sustainable Future: Sequel to the Limits to Growth*. London: Earthscan.

Mellor, M. (1992) *Breaking the Boundaries: Towards a Feminist Green Socialism*. London: Virago Press.

Mellor, M. (1997) *Feminism and Ecology*. Cambridge: Polity.

Melucci, A. (1989) *Nomads of the Present: Social Movements and Individual Needs in Contemporary Society*. London: Radius.

Melucci, A. (1996) *Changing Codes*. Cambridge: Cambridge University Press.

Merchant, C. (1980) *The Death of Nature: Women, Ecology and the Scientific Revolution*. New York: Harper and Row.

Merchant, C. (1992) *Radical Ecology: The Search for a Livable World*. London: Routledge.

Middleton., N., O'Keefe, P. and Moyo, S. (1993) *Tears of the Crocodile: From Rio to Reality in the Developing World*. London: Pluto

Midgley, M. (1983) *Animals and Why They Matter: A Journey around the Species Barrier*. Harmondsworth: Penguin.

Midgley, M. (1994) 'Bridge-building at Last', in A. Manning and J. Serpell (eds) *Animals and Human Society*. London: Routledge.

Mies, M. (1986) *Patriarchy and Accumulation on a World Scale*. London: Zed Press.

Mies, M. and Shiva, V. (1993) *Ecofeminism*. London: Zed Books.

Mies, M., Bennholdt-Thompson, V. and von Werlhof, C. (1988) *Women: The Last Colony*. London: Zed Press.

Miller, M.A.L. (1995) *The Third World in Global Environmental Politics*. Buckingham: Open University Press.

Mumford, L. (1961) *The City in History*. London: Secker and Warburg.

Murdock, J. (2001) 'Ecologising Sociology: Actor-Network Theory, Co-constructionism and the Problem of Human Exemptionalism', *Sociology* 35, 1: 111–33.

Murphy, R. (1994a) 'The Sociological Construction of Science without Nature', *Sociology* 28, 4: 957–74.

Murphy, R. (1994b) *Rationality and Nature*. Boulder, CO: Westview Press.

Naess, A. (1973) 'The Shallow and the Deep, Long-Range Ecology Movement: A Summary', *Inquiry* 16: 95–100.

Naess, A. (1984) 'Intuition, Intrinsic Value and Deep Ecology', *The Ecologist* 14: 5–6.

Naess, A. (1989) *Ecology, Community and Lifestyle: Outline of an Ecosophy*. Cambridge: Cambridge University Press.

Naess, A. (1990) 'Deep Ecology', in A. Dobson (ed.) *The Green Reader*. London: André Deutsch.

Newby, H. (1979) *Green and Pleasant Land? Social Change in Rural England*. Harmondsworth: Penguin.

Newby, H. (1991) 'One World, Two Cultures: Sociology and the Environment', *Network* 50: 1–8.

Newby, J. (1997) *The Pact for Survival*. Sydney: ABC Books.

Nicholson, L. (ed.) (1990) *Feminism/Postmodernism*. London: Routledge.

Oakley, A. (1976) *Housewife*. Harmondsworth: Penguin.

O'Brien, M. (1981) *The Politics of Reproduction*. London: Routledge and Kegan Paul.

O'Connor, J. (1989) 'Political Economy and the Ecology of Socialism and Capitalism', *Capitalism, Nature, Socialism* 1, 1: 11–38.

O'Connor, J. (1994) 'On the Misadventures of Capitalist Nature', in J. Connor (ed.) *Is Capitalism Sustainable?* London: Guilford Press.

Offe, C. (1985a) 'New Social Movements: Challenging the Boundaries of Institutional Politics', *Social Research* 52, 4: 817–68.

Offe, C. (1985b) *Disorganized Capitalism: Contemporary Transformations of Work and Politics*. Cambridge: Cambridge University Press.

Ohmae, K. (1990) *The Borderless World*. London: Collins.

Ohmae, K. (1995) *The End of the Nation State*. New York: Free Press.

O'Neill, J. (1993) *Ecology, Policy and Politics: Human Well-Being in the Natural World*. London: Routledge.

Ophuls, W. (1977) *Ecology and the Politics of Scarcity: A Prologue to the Political Theory of the Steady State*. San Francisco, CA: Freeman.

O'Riordan, T. (1981) *Environmentalism*, 2nd edn. London: Pion.

Orwell, G. (1949) *Animal Farm*. London: Secker and Warburg.

Pakulski, J. (1991) *Social Movements: The Politics of Social Protest*. Melbourne: Longman Cheshire.

Palmer, J. (ed.) (2001) *50 Key Thinkers on the Environment*. London: Routledge.

Park, R.E. (1952) *Human Communities: The City and Human Ecology*. Glencoe, IL: The Free Press.

Parsons, T. (1960) *Structure and Process in Modern Societies*. Glencoe, IL: The Free Press.

Passmore, J. (1980) *Man's Responsibility for Nature*, 2nd edn. London: Duckworth.

Pearce, F. (1991) *Green Warriors*. London: Bodley Head.

Peet, R. and Watts, M. (eds) (1996a) *Liberation Ecologies*. London: Routledge.

Peet, R. and Watts, M. (1996b) 'Liberation Ecology: Development, Sustainability, and Environment in an Age of Market Triumphalism', in R. Peet and M. Watts (eds) *Liberation Ecologies*. London: Routledge.

Pepper, D. (1984) *The Roots of Modern Environmentalism*. London: Croom Helm.

Pepper, D. (1991) *Communes and the Green Vision: Counter-Culture, Lifestyle and the New Age*. London: Green Print.

Pepper, D. (1993) *Eco-Socialism: From Deep Ecology to Social Justice*. London: Routledge.

Pepper, D. (1996) *Modern Environmentalism: An Introduction*. London: Routledge.

Plant, J. (ed.) (1989) *Healing the Wounds: The Promise of Ecofeminism*. London: Green Print.

Plumwood, V. (1988) 'Women, Humanity and Nature', *Radical Philosophy* Spring: 16–24.

Plumwood, V. (1991) *Nature, Self and Gender*. London: Routledge.

Plumwood, V. (1993) *Feminism and the Mastery of Nature*. London: Routledge.

Plumwood, V. (1994) 'The Ecopolitics Debate and the Politics of Nature', in K. Warren (ed.) *Ecological Feminism*. London: Routledge.

Porritt, J. (1986) *Seeing Green: The Politics of Ecology Explained*. Oxford: Blackwell.

Porritt, J. and Winner, M. (1988) *The Coming of the Greens*. London: Fontana.

Porter, G. and Brown, J. (1991) *Global Environmental Politics: Dilemmas in World Politics*. Boulder, CO: Westview Press.

Pulzer, P. (1995) *German Politics, 1945–1995*. Oxford: Oxford University Press.

Pye, L. (1966) *Aspects of Political Development*. Boston, MA: Little Brown.

Randall, V. and Theobald, C. (1985) *Political Change and Underdevelopment*. London: Macmillan.

Rangan, H. (1996) 'From Chipko to Uttaranchal: Development, environment and social protest in the Garhwal Himalayas, India', in R. Peet and M. Watts (eds) *Liberation Ecologies*. London: Routledge.

Ratzan, S.C. (ed.) (1998) *The Mad Cow Crisis: Health and the Public Good*. London: UCL Press.

Redclift. M. (1984) *Development and the Environmental Crisis: Red or Green Alternatives?* London: Methuen.

Redclift. M. (1987) *Sustainable Development: Exploring the Contradictions*. London: Methuen.

Redclift, M. and Benton, T. (eds) (1994) *Social Theory and the Global Environment*. London: Routledge.

Regan, T. (1988) *The Case for Animal Rights*. London: Routledge.

Richardson, D. and Rootes, C. (eds) (1995) *The Green Challenge*. London: Routledge.

Rifkin, J. (1994) *Beyond Beef: The Rise and Fall of Cattle Culture*. London: Thorsons.

Ritvo, H. (1987) *The Animal Estate: The English and Other Creatures in the Victorian Age*. Cambridge, MA: Harvard University Press.

Ritvo, H. (1994) 'Animals in Nineteenth-Century Britain', in A. Manning and J. Serpell (eds) *Animals and Human Society*. London: Routledge.

Ritzer, G. (1996) *The McDonaldization of Society*, revised edn. London: Sage/Pine Forge Press.

Robertson, R. (1992) *Globalization: Social Theory and Global Culture*. London: Sage.

Rollin, B.E. (1981) *Animal Rights and Human Morality*. New York: Prometheus Books.

Rootes, C. (ed.) (1990) *Environmental Movements: Local, National and Global*. London: Frank Cass.

Rose, H. (1994) *Love, Power and Knowledge*. Cambridge: Polity.

Rosenau, J. (1990) *Turbulence in World Politics*. Brighton: Harvester Weatsheaf.

Rosenau, J. (1997) *Along the Domestic–Foreign Frontier*. Cambridge: Cambridge University Press.

Rosenau, J. (1998) 'Government and Democracy in a Globalizing World', in D. Archibugi, D. Held and M. Kohler (eds) *Re-imagining Political Community*. Cambridge: Polity Press.

Rostow, W. (1962) *The Stages of Economic Growth: A Non-communist Manifesto*. Cambridge: Cambridge University Press.

Roszak, T. (1979) *Person/Planet: The Creative Disintegration of Industrial Society*. London: Victor Gollancz.

Roszak, T. (1989) *Where the Wasteland Ends*. Berkeley: Celestial Arts.

Roszak, T. (1992) *The Voice of the Earth: An Exploration in Ecopsychology*. New York: Simon and Schuster.

Roxborough, I. (1979) *Theories of Underdevelopment*. London: Macmillan.

Rüdig, W., Bennie, L. and Franklin, M. (1991) *Green Party Members: A Profile*. Glasgow: Delta.

Ryder, R.D. (1989) *Animal Revolution: Changing Attitudes Towards Speciesism*. Oxford: Blackwell.

Ryder, R.D. (1992) *Painism; Ethics, Animal Rights and Environmentalism*. Cardiff: University of Wales College of Cardiff.

Ryle, M. (1988) *Ecology and Socialism*. London: Century Hutchinson.

Sachs, W. (ed.) (1993) *Global Ecology*. London: Zed.

Said, E. (1978) *Orientalism*. New York: Pantheon.

Sale, K. (1980) *Human Scale*. New York: Coward, Cann and Geoghegan.

Sale, K. (1985) *Dwellers in the Land: A Bioregional Vision*. San Francisco, CA: Sierra Club Books.

Salleh, A.K. (1984) 'Deeper than Deep Ecology: The Ecofeminist Connection', *Environmental Ethics* 6: 339–45.

Salleh, A. (1997) *Ecofeminism as Politics: Nature, Marx and the Postmodern*. London: Zed Press.

Salmon, P.W. and Salmon, I.M. (1983) 'Who Owns Who? Psychological Research into the Human–Pet Bond in Australia', in A.H. Katcher and A.M. Beck (eds) *New Perspectives in Our Lives with Companion Animals*. Philadelphia: University of Pennsylvania Press.

Salt, H. (1980) *Animal Rights Considered in Relation to Social Progress*. London: Centaur. (First published 1892.)

Sayer, A. (1984) *Method in Social Science: A Realist Approach*. London: Hutchinson.

Schnailberg, A. (1980) *The Environment: from Surplus to Scarcity*. Oxford: Oxford University Press.

Schnailberg, A. and Gould, K. (1994) *Environment and Society: The Enduring Conflict*. New York: St Martins Press.

Schumacher, E.F. (1973) *Small is Beautiful: Economics as if People Mattered*. London: Sphere.

Scott, A. (1990) *Ideology and the New Social Movements*. London: Unwin Hyman.

Seabrooke, J. (1986) 'The Inner City Environment: Making the Connections', in J. Weston (ed.) *Red and Green*. London: Pluto.

Segal, L. (1987) *Is the Future Female? Troubled Thoughts on Contemporary Feminism*. London: Virago.

Sen, G. and Grown, C. (1987) *Development, Crises and Alternative Visions*. New York: Monthly Review Press.

Serpell, J. (1986) *In the Company of Animals*. Oxford: Blackwell.

Serpell, J. (1994) 'Attitudes, Knowledge and Behaviour Towards Wildlife among the Industrial Superpowers: The United States, Japan and Germany', in A. Manning and J. Serpell (eds) *Animals and Human Society – Changing Perspectives*. London: Routledge.

Serpell, J. and Paul, E. (1994) 'Pets and Positive Attitudes to Animals', in A. Manning and J. Serpell (eds) *Animals and Human Society – Changing Perspectives*. London: Routledge.

Sessions, G. (ed.) (1995) *Deep Ecology for the Twenty-First Century*. Boston, MA: Shambhala Press.

Shiva, V. (1988) *Staying Alive: Women, Ecology and Development*. London: Zed Press.

Shiva, V. (1993) *Monocultures of the Mind*. London: Zed Press.

Shiva, V. (1998) *Biopiracy: The Plunder of Nature and Knowledge*. Dartington: Green Books.

Shoard, M. (1980) *The Theft of the Countryside*. London: Temple Smith.

Shoard, M. (1987) *This Land is Our Land: The Struggle for Britain's Countryside*. London: Paladin.

Short, J.R. (1989) *The Humane City: Cities as if People Really Mattered*. Oxford: Basil Blackwell.

Short, J.R. (1991) *Imagined Country: Environment, Culture and Society*. London: Routledge.

Simmel, G. (1991) 'The Metropolis and Mental Life', in G. Simmel, *On Individuality and Social Forms: Selected Writings*, ed. D.N. Levine. London: University of Chicago Press.

Simmons, I.G. (1993) *Environmental History*. Oxford: Blackwell.

Singer, P. (1979) *Practical Ethics*. Boston, MA: Cambridge University Press.

Singer, P. (1981) *The Expanding Circle: Ethics and Sociobiology*. New York: Farrar, Strauss and Giroux.

Singer, P. (ed.) (1985) *In Defence of Animals*. Oxford: Blackwell.

Singer, P. (1990) *Animal Liberation*, 2nd edn. New York: Avon Books.

Sklair, L. (1991) *Sociology of the Global System*. Hemel Hempstead: Harvester Wheatsheaf.

Sklair, L. (1994) 'Global Sociology and Global Environmental Change', in M. Redclift and T. Benton (ed.) *Social Theory and Global Environmental Change*. London: Routledge.

Smelser, N. (1962) *Theory of Collective Behaviour*. London: Routledge and Kegan Paul.

Sontheimer, S. (ed.) (1991) *Women and the Environment: A Reader*. London: Earthscan.

Soper, K. (1995) *What is Nature?* Oxford: Blackwell.

Soper, K. (1996a) 'Greening the Prometheus: Marxism and Ecology', in T. Benton (ed.) *The Greening of Marxism*. London: Guilford Press.

Soper, K. (1996b) 'Feminism, Ecosocialism and the Conceptualization of Nature', in T. Benton (ed.) *The Greening of Marxism*. London: Guilford Press.

Spelman, E. (1988) *Inessential Woman*. London: The Women's Press.

Spiegal, M. (1988) *The Dreaded Comparison: Human and Animal Slavery*. London: Routledge.

Spretnak, C. (1982) *The Politics of Women's Spirituality*. New York: Doubleday.

Spretnak, C. (1985) 'The Spiritual Dimension of Green Politics', in C. Spretnak and F. Capra (eds) *Green Politics*. Glasgow: Paladin.

Spretnak, C. (1990) 'Ecofeminism: Our Roots and Flowering', in I. Diamond and G. Orenstein (eds) *Reweaving the World*. San Francisco, CA: Sierra Club Books.

Spretnak, C. (1991) *States of Grace*. New York: HarperCollins.

Spretnak, C. and Capra, F. (1985) *Green Politics: The Global Promise*. Glasgow: Paladin.

Starhawk (1979) *The Spiral Dance: A Rebirth of the Ancient Religion of the Great Goddess*, revised edn, 1989. San Francisco, CA: Harper and Row.

Starhawk (1982) *Dreaming the Dark: Magic, Sex and Politics*. Boston, MA: Beacon Press.

Starhawk (1987) *Truth or Dare: Encounters with Power, Authority and Mystery*. San Francisco, MA: Harper and Row.

Starhawk (1990) 'Power, Authority, Mystery: Eco-feminism and Earth-based Spirituality', in I. Diamond and G. Orenstein, (eds) *Reweaving the World*. San Francisco, CA: Sierra Club Books.

Tester, K. (1991) *Animals and Society: The Humanity of Animal Rights*. London: Routledge.

Thomas, K. (1983) *Man and the Natural World: Changing Attitudes in England 1500–1800*. London: Allen Lane.

Thompson, E.P. (1968) *The Making of the English Working Class*. Harmondsworth: Penguin.

Tilly, C. (1978) *From Mobilization to Revolution*. Reading, MA: Addison Wesley.

Tobias, M. (ed.) (1985) *Deep Ecology*. San Diego, CA: Avant.

Tokar, B. (1987) *The Green Alternative*. San Pedro, CA: R. and E. Miles.

Tokar, B. (1988) 'Social Ecology, Eeep Ecology and the Future of Green Thought', *The Ecologist* 18: 4–5.

Tönnies, F. (1963) *Community and Society*. New York: Harper and Row.

Touraine, A. (1971) *Post-Industrial Society*. New York: Random House.

Touraine, A. (1981) *The Voice and the Eye: An Analysis of Social Movements*. Cambridge: Cambridge University Press.

Trainer, T. (1985) *Abandon Affluence!* London: Zed Books.

Twigg, J. (1983) 'Vegetarianism and the Meanings of Meat', in A. Murcott (ed.) *The Sociology of Food and Eating*. Aldershot: Gower.

Urry, J. (1995) *Consuming Places*. London: Routledge.

Urry, J. (2000) *Sociology Beyond Societies: Mobilities for the Twenty-first Century*. London: Routledge.

Wall, D. (1999) *Earth First! and the Anti-Roads Movement*. London: Routledge.

Wallerstein, I. (1974) *The Modern World System: Capitalist Agriculture and the Origins of the European World Economy in the Sixteenth Century*. London: Academic Press.

Wallerstein, I. (1979) *The Capitalist World Economy*. Cambridge: Cambridge University Press.

Wallerstein, I. (1990) 'Societal Development, or Development of the World System?' in M. Albrow and E. King (eds) *Globalization, Knowledge and Society*. London: Sage.

Ward, B. and Dubos, R. (1972) *Only One Earth: The Care and Maintenance of a Small Planet*. London: Penguin/André Deutsch.

Warde, A. (1997) *Consumption, Food and Taste*. London: Sage.

Warren, B. (1980) *Imperialism, Pioneer of Capitalism*. London: Verso.

Warren, K. (1987) 'Feminism and Ecology: Making Connections', *Environmental Ethics* 9: 3–20.

Warren, K. (1990) 'The Power and the Promise of Ecological Feminism', *Environmental Ethics* 12: 125–46.

Warren, K. (ed.) (1994) *Ecological Feminism*. London: Routledge.

Warren, K. (2000) *Ecofeminist Philosophy*. Savage, MD: Rowland and Littlefield.

Waters, J. (1995) *Globalization*. London: Routledge.

Weber, M. (1938) *The Protestant Ethic and the Spirit of Capitalism*. London: Unwin.

Weber, M. (1964) *The Theory of Social and Economic Organization*. London: Macmillan.

Weiss, L. (1998) *State Capacity: Governing the Economy in the Global Era*. Cambridge: Polity Press.

Weston, J. (ed.) (1986) *Red and Green: A New Politics of the Environment*. London: Pluto.

Williams, R. (1973) *The Country and the City*. London: Chatto and Windus.

Wilson, E. (1991) *The Sphinx in the City: Urban Life, the Control of Disorder, and Women*. London: Virago.

Wirth, L. (1938) 'Urbanism as a Way of Life', in A.J. Reiss Jr (ed.) *Louis Wirth on Cities and Social Life*. Chicago: University of Chicago Press.

Wynne, B. (1992) 'Risk and Social Learning: Reification to Engagement', in S. Krimsky and D. Golding (eds) *Social Theories of Risk*. Westport, CT: Praeger.

Wynne, B. (1996) 'May the Sheep Safely Graze? A Reflexive View of the Expert–Lay Divide, in S. Lash, B. Szerzynski and B. Wynne, *Environment and Modernity*. London: Sage.

Yearley, S. (1992) *The Green Case: A Sociology of Environmental Issues, Arguments and Politics*. London: Routledge.

Yearley, S. (1994) 'Social Movements and Environmental Change', in M. Redclift and T. Benton (eds) *Social Theory and Global Environmental Change*. London: Routledge.

Yearley, S. (1996) *Sociology, Environmentalism, Globalization*. London: Sage.

Young, J. (1989) *Postenvironmentalism*. London: Belhaven Press.

Young, M. and Wilmott, P. (1960) *Family and Class in a London Suburb*. Harmondsworth: Penguin.

Young, M. and Willmott, P. (1962) *Family and Kinship in East London*, revised edn. Harmondsworth: Penguin.

Zirakzadeh, C. (1997) *Social Movements in Politics: A Comparative Study*. London: Longman.

 **Index**

OTHER TITLES FROM THE ROUTLEDGE INTRODUCTIONS TO ENVIRONMENT SERIES

Environment and Social Theory
John Barry

Routledge

Hb: 0–415–17269–1
Pb: 0–415–17270–5

Environment and Politics
2nd edition
Timothy Doyle and Doug McEachern

Routledge

Hb: 0–415–21772–5
Pb: 0–415–21773–3

Environmental Movements
Timothy Doyle

Routledge

Hb: 0–415–19068–1
Pb: 0–415–19069–x

Environment and Law
David Wilkinson

Routledge

Hb: 0–415–21567–6
Pb: 0–415–21568–4

Representing the Environment
John Gold and George Revill

Routledge

Hb: 0–415–14589–9
Pb: 0–415–14590–2

Information and ordering details

For price availability and ordering visit our website www.tandf.co.uk
Subject Web Address **www.geographyarena.com**
Alternatively our books are available from all good bookshops.

Learning Resources